UNWEAVING THE RAINBOW

UNWEAVING THE RAINBOW

SCIENCE, DELUSION AND THE APPETITE FOR WONDER

RICHARD DAWKINS

ALLEN LANE
THE PENGUIN PRESS

ALLEN LANE
THE PENGUIN PRESS

Published by the Penguin Group
Penguin Books Ltd, 27 Wrights Lane, London W8 5TZ, England
Penguin Putnam Inc., 375 Hudson Street, New York, New York 10014, USA
Penguin Books Australia Ltd, Ringwood, Victoria, Australia
Penguin Books Canada Ltd, 10 Alcorn Avenue, Toronto, Ontario, Canada M4V 3B2
Penguin Books (NZ) Ltd, Private Bag 102902, NSMC, Auckland, New Zealand

Penguin Books Ltd, Registered Offices: Harmondsworth, Middlesex, England

First published 1998
1 3 5 7 9 10 8 6 4 2

Acknowledgement is made to the following for permission to use extracts from: 'Passion' by Kathleen Raine, reprinted by permission of the author; 'Agamemnon's Tomb' by Sacheverell Sitwell, reprinted by permission of David Higham Associates; 'Unpredictable But Providential (for Loren Eiseley)' by W. H. Auden, reprinted by permission of Faber and Faber Ltd

Set in 12.25/15.5 pt PostScript Monotype Walbaum
Typeset by Rowland Phototypesetting Ltd, Bury St Edmunds, Suffolk
Printed in England by Clays Ltd, St Ives plc

A CIP catalogue record for this book is available from the British Library

ISBN 0-713-99214-X

For Lalla

CONTENTS

	PREFACE	ix
1	*THE ANAESTHETIC OF FAMILIARITY*	1
2	*DRAWING ROOM OF DUKES*	15
3	*BARCODES IN THE STARS*	38
4	*BARCODES ON THE AIR*	66
5	*BARCODES AT THE BAR*	83
6	*HOODWINK'D WITH FAERY FANCY*	114
7	*UNWEAVING THE UNCANNY*	145
8	*HUGE CLOUDY SYMBOLS OF A HIGH ROMANCE*	180
9	*THE SELFISH COOPERATOR*	210
10	*THE GENETIC BOOK OF THE DEAD*	235
11	*REWEAVING THE WORLD*	257
12	*THE BALLOON OF THE MIND*	286
	SELECTED BIBLIOGRAPHY	314
	INDEX	325

PREFACE

A foreign publisher of my first book confessed that he could not sleep for three nights after reading it, so troubled was he by what he saw as its cold, bleak message. Others have asked me how I can bear to get up in the mornings. A teacher from a distant country wrote to me reproachfully that a pupil had come to him in tears after reading the same book, because it had persuaded her that life was empty and purposeless. He advised her not to show the book to any of her friends, for fear of contaminating them with the same nihilistic pessimism. Similar accusations of barren desolation, of promoting an arid and joyless message, are frequently flung at science in general, and it is easy for scientists to play up to them. My colleague Peter Atkins begins his book *The Second Law* (1984) in this vein:

We are the children of chaos, and the deep structure of change is decay. At root, there is only corruption, and the unstemmable tide of chaos. Gone is purpose; all that is left is direction. This is the bleakness we have to accept as we peer deeply and dispassionately into the heart of the Universe.

But such very proper purging of saccharine false purpose; such laudable tough-mindedness in the debunking of cosmic sentimentality must not be confused with a loss of personal hope. Presumably there is indeed no purpose in the ultimate fate of the cosmos, but do any of us really tie our life's hopes to the ultimate fate of the cosmos anyway? Of course we don't; not if we are sane. Our lives are ruled by all sorts of closer, warmer, human ambitions and perceptions. To accuse science of robbing life of the warmth that makes it worth living is so preposterously mistaken, so diametrically opposite to my own feelings and those of most working scientists, I am almost driven to the despair of which I am wrongly suspected. But in this book I shall try a more positive response, appealing to the sense of wonder in science because it is so sad to think what these complainers and naysayers are *missing*. This is one of the things that the late Carl Sagan did so well, and for which he is sadly missed. The feeling of awed wonder that science can give us is one of the highest experiences of which the human psyche is capable. It is a deep aesthetic passion to rank with the finest that music and poetry can deliver. It is truly one of the things that makes life worth living and it does so, if anything, more effectively if it convinces us that the time we have for living it is finite.

My title is from Keats, who believed that Newton had destroyed all the poetry of the rainbow by reducing it to the prismatic colours. Keats could hardly have been more wrong, and my aim is to guide all who are tempted by a similar view towards the opposite conclusion. Science is, or ought to be, the inspiration for great poetry, but I do not have the talent to clinch the argument by demonstration and must depend, instead, on more prosaic persuasion. A couple of the chapter titles are borrowed from Keats; readers may also spot the occasional half-quotation or allusion lacing the text from him (as well as others). They are there as a tribute to his sensitive genius. Keats was a more likeable character than Newton and his shade was one of the imaginary referees looking over my shoulder as I wrote.

Newton's unweaving of the rainbow led on to spectroscopy, which has proved the key to much of what we know today about the cosmos. And the heart of any poet worthy of the title Romantic could not fail to leap up if he beheld the universe of Einstein, Hubble and Hawking. We read its nature through Fraunhofer lines – 'Barcodes in the Stars' – and their shifts along the spectrum. The image of barcodes carries us on to the very different, but equally intriguing, realms of sound ('Barcodes on the Air'); and then DNA fingerprinting ('Barcodes at the Bar'), which offers the opportunity to reflect on other aspects of the role of science in society.

In what I call the Delusion section of the book, 'Hoodwink'd with Faery Fancy' and 'Unweaving the Uncanny', I turn to those ordinary superstitious folk who, less exalted than poets defending rainbows, revel in mystery and feel cheated if it is explained. They are the ones who love a good ghost story, whose mind leaps to poltergeists or miracles whenever something even faintly odd happens. They never lose an opportunity to quote Hamlet's

There are more things in heaven and earth, Horatio,
Than are dreamt of in your philosophy.

and the scientist's response ('Yes, but we're working on it') strikes no chord with them. For them, to explain away a good mystery is to be a killjoy, just as some Romantic poets thought about Newton's explaining of the rainbow.

Michael Shermer, editor of *Skeptic* magazine, tells a salutary story of an occasion when he publicly debunked a famous television spiritualist. The man was doing ordinary conjuring tricks and duping people into thinking he was communicating with dead spirits. But instead of being hostile to the now-unmasked charlatan, the audience turned on the debunker and supported a woman who accused him of 'inappropriate' behaviour because he destroyed people's illusions. You'd think she'd have been grateful for having the wool pulled *off* her eyes, but apparently she preferred it firmly over them. I believe that an orderly universe, one indifferent to human preoccupations, in which everything has an explanation even if we still have a long way to go before we find it, is a more beautiful, more wonderful place than a universe tricked out with capricious, *ad hoc* magic.

Paranormalism could be called an abuse of the legitimate sense of poetic wonder which true science ought to be feeding. A different threat comes from what may be called bad poetry. The chapter on 'Huge Cloudy Symbols of a High Romance' warns against seduction by bad poetic science; against the allure of misleading rhetoric. By way of example, I look at a particular contributor to my own field whose imaginative writing has given him a disproportionate – and I believe unfortunate – influence on American understanding of evolution. But the dominant thrust of the book is in favour of good poetic science, by which I don't, of course, mean science written in verse but science inspired by a poetic sense of wonder.

The last four chapters attempt, with respect to four different but interrelated topics, to hint at what might be done by poetically inspired scientists more talented than I am. Genes, however 'selfish', must also be 'cooperative' – in an Adam Smithian sense (which is why the chapter 'The Selfish Cooperator' opens with a quotation from Adam Smith, though admittedly not on this topic but on wonder itself). The genes of a species can be thought of as a description of ancestral worlds, a 'Genetic Book of the Dead'. In a similar way, the brain 'reweaves the world', constructing a kind of 'virtual reality' continuously updated in the head. In 'The Balloon of the Mind' I speculate on the origins of our own species' most unique features and return, finally, to wonder at the poetic impulse itself and the part it may have played in our evolution.

Computer software is driving a new renaissance, and some of its creative geniuses are benefactors and simultaneously renaissance men in their own right. In 1995, Charles Simonyi of Microsoft endowed a new professorship of Public Understanding of Science at the University of Oxford, and I was appointed its first holder. I am grateful to Dr Simonyi, most obviously for his far-sighted generosity towards a university with which he had no previous connection, but also for his imaginative vision of science and how it should be communicated. This was beautifully expressed in his written statement to the Oxford of the future (his endowment is in perpetuity, yet he characteristically eschews the wary meanness of lawyer language) and we have discussed these matters from time to time since becoming friends after my appointment. *Unweaving the Rainbow* could be seen as my contribution to the conversation, and as my inaugural statement as Simonyi Professor. And if 'inaugural' sounds a little unbecoming after two years in the job, I may perhaps take a liberty and quote Keats again:

> *By this, friend Charles, you may full plainly see*
> *Why I have never penn'd a line to thee:*
> *Because my thoughts were never free, and clear,*
> *And little fit to please a classic ear.*

Nevertheless, it is in the nature of a book that it takes longer to produce than a newspaper article or a lecture. During its gestation this one has spun off a few of both, and broadcasts as well. I must acknowledge these now, in case any readers recognize the odd paragraph here and there. I first publicly used the title 'Unweaving the Rainbow', and the theme of Keats's irreverence towards Newton, when I was invited to give the C. P. Snow Lecture for 1997 by Christ's College, Cambridge, Snow's old college. Although I have not explicitly taken up his theme of *The Two Cultures*, it is obviously relevant. Even more so is *The Third Culture* of John Brockman, who has been helpful, too, in a quite different role, as my literary agent. The subtitle 'Science, Delusion and the Appetite for Wonder' was the title of my Richard Dimbleby Lecture, 1996. Some paragraphs from an earlier draft of this book appeared in that BBC televised lecture. Also in 1996, I presented a one-hour television documentary on Channel Four, *Break the Science Barrier*. This was on the theme of science in the culture, and some of the background ideas, developed in discussions with John Gau, the producer, and Simon Raikes, the director, have influenced this book. In 1998 I incorporated some passages of the book in my lecture in the *Sounding the Century* series broadcast by BBC Radio 3 from the Queen

Elizabeth Hall, London. (I thank my wife for my lecture's title, 'Science and Sensibility', and don't quite know what to make of the fact that it has already been plagiarized in, of all places, a supermarket magazine.) I also have used paragraphs from the book in articles commissioned by the *Independent*, the *Sunday Times* and the *Observer*. When I was honoured with the 1997 International Cosmos Prize, I chose the title 'The Selfish Cooperator' for my prize lecture, given in both Tokyo and Osaka. Parts of the lecture have been reworked and expanded in chapter 9, which has the same title.

The book has benefited greatly from constructive criticisms of an earlier draft by Michael Rodgers, John Catalano and Lord Birkett. Michael Birkett has become my ideal intelligent layman. His scholarly wit makes his critical comments a pleasure to read in their own right. Michael Rodgers was the editor of my first three books and, by my wish and his generosity, he has also played an important role in the last three as well. I would like to thank John Catalano, not just for his helpful comments on the book but for http://www.spacelab.net/~catalj/home.html, whose excellence – which has nothing whatever to do with me – will be apparent to all who go there. Stefan McGrath and John Radziewicz, editors at Penguin and Houghton Mifflin respectively, gave patient encouragement and literate advice which I greatly valued. Sally Holloway worked tirelessly and cheerfully on the final copy-editing. Thanks also to Ingrid Thomas, Bridget Muskett, James Randi, Nicholas Davies, Daniel Dennett, Mark Ridley, Alan Grafen, Juliet Dawkins, Anthony Nuttall and John Batchelor.

My wife, Lalla Ward, has criticized every chapter a dozen times in various drafts, and with every reading I have benefited from her sensitive actor's ear for language and its cadences. Whenever I had doubts, she believed in the book. Her vision held it together, and I wouldn't have finished it without her help and encouragement. I dedicate it to her.

1

THE ANAESTHETIC OF FAMILIARITY

To live at all is miracle enough.

MERVYN PEAKE,
The Glassblower (1950)

We are going to die, and that makes us the lucky ones. Most people are never going to die because they are never going to be born. The potential people who could have been here in my place but who will in fact never see the light of day outnumber the sand grains of Arabia. Certainly those unborn ghosts include greater poets than Keats, scientists greater than Newton. We know this because the set of possible people allowed by our DNA so massively exceeds the set of actual people. In the teeth of these stupefying odds it is you and I, in our ordinariness, that are here.

Moralists and theologians place great weight upon the moment of conception, seeing it as the instant at which the soul comes into existence. If, like me, you are unmoved by such talk, you still must regard a particular instant, nine months before your birth, as the most decisive event in your personal fortunes. It is the moment at which your consciousness suddenly became trillions of times more foreseeable than it was a split second before. To be sure, the embryonic you that came into existence still had plenty of hurdles to leap. Most conceptuses end in early abortion before their mother even knew they were there, and we are all lucky not to have done so. Also, there is more to personal identity than genes, as identical twins (who separate after the moment of fertilization) show us.

Nevertheless, the instant at which a particular spermatozoon penetrated a particular egg was, in your private hindsight, a moment of dizzying singularity. It was then that the odds against your becoming a person dropped from astronomical to single figures.

The lottery starts before we are conceived. Your parents had to meet, and the conception of each was as improbable as your own. And so on back, through your four grandparents and eight great grandparents, back to where it doesn't bear thinking about. Desmond Morris opens his autobiography, *Animal Days* (1979), in characteristically arresting vein:

Napoleon started it all. If it weren't for him, I might not be sitting here now writing these words . . . for it was one of his cannonballs, fired in the Peninsular War, that shot off the arm of my great-great-grandfather, James Morris, and altered the whole course of my family history.

Morris tells how his ancestor's enforced change of career had various knock-on effects culminating in his own interest in natural history. But he really needn't have bothered. There's no 'might' about it. *Of course* he owes his very existence to Napoleon. So do I and so do you. Napoleon didn't have to shoot off James Morris's arm in order to seal young Desmond's fate, and yours and mine, too. Not just Napoleon but the humblest medieval peasant had only to sneeze in order to affect something which changed something else which, after a long chain reaction, led to the consequence that one of your would-be ancestors failed to be your ancestor and became somebody else's instead. I'm not talking about 'chaos theory', or the equally trendy 'complexity theory', but just about the ordinary statistics of causation. The thread of historical events by which our existence hangs is wincingly tenuous.

When compared with the stretch of time unknown to us, O king, the present life of men on earth is like the flight of a single sparrow through the hall where, in winter, you sit with your captains and ministers. Entering at one door and leaving by another, while it is inside it is untouched by the wintry storm; but this brief interval of calm is over in a moment, and it returns to the winter whence it came, vanishing from your sight. Man's life is similar; and of what follows it, or what went before, we are utterly ignorant.

THE VENERABLE BEDE,
A History of the English Church and People (731)

This is another respect in which we are lucky. The universe is older than a hundred million centuries. Within a comparable time the sun will swell to a red giant and engulf the earth. Every century of hundreds of millions has been in its time, or will be when its time comes, 'the present century'. Interestingly, some physicists don't like the idea of a 'moving present', regarding it as a subjective phenomenon for which they find no house room in their equations. But it is a subjective argument I am making. How it feels to me, and I guess to you as well, is that the present moves from the past to the future, like a tiny spotlight, inching its way along a gigantic ruler of time. Everything behind the spotlight is in darkness, the darkness of the dead past. Everything ahead of the spotlight is in the darkness of the unknown future. The odds of your century being the one in the spotlight are the same as the odds that a penny, tossed down at random, will land on a particular ant crawling somewhere along the road from New York to San Francisco. In other words, it is overwhelmingly probable that you are dead.

In spite of these odds, you will notice that you are, as a matter of fact, alive. People whom the spotlight has already passed over, and people whom the spotlight has not reached, are in no position to read a book. I am equally lucky to be in a position to write one, although I may not be when you read these words. Indeed, I rather hope that I shall be dead when you do. Don't misunderstand me. I love life and hope to go on for a long time yet, but any author wants his works to reach the largest possible readership. Since the total future population is likely to outnumber my contemporaries by a large margin, I cannot but aspire to be dead when you see these words. Facetiously seen, it turns out to be no more than a hope that my book will not soon go out of print. But what I see as I write is that I am lucky to be alive and so are you.

We live on a planet that is all but perfect for our kind of life: not too warm and not too cold, basking in kindly sunshine, softly watered; a gently spinning, green and gold harvest festival of a planet. Yes, and alas, there are deserts and slums; there is starvation and racking misery to be found. But take a look at the competition. Compared with most planets this is paradise, and parts of earth are still paradise by any standards. What are the odds that a planet picked at random would have these complaisant properties? Even the most optimistic calculation would put it at less than one in a million.

Imagine a spaceship full of sleeping explorers, deep-frozen would-be colonists of some distant world. Perhaps the ship is on a forlorn mission to save the species before an unstoppable comet, like the one that killed the dinosaurs, hits the home planet. The voyagers go into the deep-freeze soberly reckoning the odds against their spaceship's ever chancing upon a planet friendly to life. If one in a million planets is suitable at best, and it takes centuries to travel from each star to the next, the spaceship is pathetically unlikely to find a tolerable, let alone safe, haven for its sleeping cargo.

But imagine that the ship's robot pilot turns out to be unthinkably lucky. After millions of years the ship does find a planet capable of sustaining life: a planet of equable temperature, bathed in warm starshine, refreshed by oxygen and water. The passengers, Rip van Winkels, wake stumbling into the light. After a million years of sleep, here is a whole new fertile globe, a lush planet of warm pastures, sparkling streams and waterfalls, a world bountiful with creatures, darting through alien green felicity. Our travellers walk entranced, stupefied, unable to believe their unaccustomed senses or their luck.

As I said, the story asks for too much luck; it would never happen. And yet, isn't that what *has* happened to each one of us? We *have* woken after hundreds of millions of years asleep, defying astronomical odds. Admittedly we didn't arrive by spaceship, we arrived by being born, and we didn't burst conscious into the world but accumulated awareness gradually through babyhood. The fact that we slowly apprehend our world, rather than suddenly discover it, should not subtract from its wonder.

Of course I am playing tricks with the idea of luck, putting the cart before the horse. It is no accident that our kind of life finds itself on a planet whose temperature, rainfall and everything else are exactly right. If the planet were suitable for another kind of life, it is that other kind of life that would have evolved here. But we as individuals are still hugely blessed. Privileged, and not just privileged to enjoy our planet. More, we are granted the opportunity to understand why our eyes are open, and why they see what they do, in the short time before they close for ever.

Here, it seems to me, lies the best answer to those petty-minded scrooges who are always asking what is the *use* of science. In one of those mythic remarks of uncertain authorship, Michael Faraday is alleged to have been asked what was the use of science. 'Sir,' Faraday replied. 'Of what use is a new-born child?' The obvious thing for Faraday (or Benjamin Franklin, or whoever it was) to have meant was that a baby might be no use for anything at present, but it has great potential for the future. I now like to think that he meant something else, too: What is the use of bringing a baby into the world if the only thing it does with its life is just work to go on living? If everything is judged by how 'useful' it is – useful for staying alive, that is – we are left facing a futile circularity. There must be some added value. At least a part of life should be devoted to *living* that life, not just working to stop it ending. This is how we rightly justify spending taxpayers' money on the arts. It is one of the justifications properly offered for conserving rare species and beautiful buildings. It is how we answer those barbarians who think that wild elephants and historic houses should be preserved only if they 'pay their way'. And science is the same. Of course science pays its way; of course it is useful. But that is not *all* it is.

After sleeping through a hundred million centuries we have finally opened our eyes on a sumptuous planet, sparkling with colour, bountiful with life. Within decades we must close our eyes again. Isn't it a noble, an enlightened way of spending our brief time in the sun, to work at understanding the universe and how we have come to wake up in it? This is how I answer when I am asked – as I am surprisingly often – why I bother to get up in the mornings. To put it the other way round, isn't it sad to go to your grave without ever wondering why you were born? Who, with such a thought, would not spring from bed, eager to resume discovering the world and rejoicing to be a part of it?

The poet Kathleen Raine, who read Natural Sciences at Cambridge, specializing in Biology, found related solace as a young woman unhappy in love and desperate for relief from heartbreak:

Then the sky spoke to me in language clear,
familiar as the heart, than love more near.
The sky said to my soul, 'You have what you desire!

'Know now that you are born along with these
clouds, winds, and stars, and ever-moving seas
and forest dwellers. This your nature is.

'Lift up your heart again without fear,
sleep in the tomb, or breathe the living air,
this world you with the flower and with the tiger share.'

'Passion' (1943)

There is an anaesthetic of familiarity, a sedative of ordinariness, which dulls the senses and hides the wonder of existence. For those of us not gifted in poetry, it is at least worth while from time to time making an effort to shake off the anaesthetic. What is the best way of countering the sluggish habituation brought about by our gradual crawl from babyhood? We can't actually fly to another planet. But we can recapture that sense of having just tumbled out to life on a new world by looking at our own world in unfamiliar ways. It's tempting to use an easy example like a rose or a butterfly, but let's go straight for the alien deep end. I remember attending a lecture, years ago, by a biologist working on octopuses, and their relatives the squids and cuttlefish. He began by explaining his fascination with these animals. 'You see,' he said, 'they are the Martians.' Have you ever watched a squid change colour?

Television images are sometimes displayed on giant LED (Light Emitting Diode) hoardings. Instead of a fluorescent screen with an electron beam scanning side to side over it, the LED screen is a large array of tiny glowing lights, independently controllable. The lights are individually brightened or dimmed so that, from a distance, the whole matrix shimmers with moving pictures. The skin of a squid behaves like an LED screen. Instead of lights, squid skin is packed with thousands of tiny bags filled with ink. Each of these ink bags has miniature private muscles to squeeze it. With a puppet string leading to each one of these separate muscles, the squid's nervous system can control the shape, and hence the visibility, of each ink sac.

In theory, if you wire-tapped the nerves leading to the separate ink pixels and stimulated them electrically via a computer, you could play out Charlie Chaplin movies on the squid's skin. The squid doesn't do that, but its brain does control the wires with precision and speed, and the skinflicks that it shows are spectacular. Waves of colour chase across the surface like clouds in a speeded-up film; ripples and eddies race over the living screen. The animal signals its changing emotions in quick time: dark brown one second, blanching ghostly white the next, rapidly modulating interwoven patterns of stipples and stripes. When it comes to changing colour, by comparison chameleons are amateurs at the game.

The American neurobiologist William Calvin is one of those thinking hard today about what thinking itself really is. He emphasizes, as others have done before, the idea that thoughts do not reside in particular places in the brain but are shifting patterns of activity over its surface, units which recruit neighbouring units into populations becoming the same thought, competing in Darwinian fashion with rival populations thinking alternative thoughts. We don't see these shifting patterns, but presumably we would if neurones lit up when active. The cortex of the brain, I realize, might then look like a squid's body surface. Does a squid think with its skin? When a squid suddenly changes its colour pattern, we suppose it to be a manifestation of mood change, for signalling to another squid. A shift in colour announces that the squid has switched from an aggressive mood, say, to a fearful one. It is natural to presume that the change in mood took place in the brain, and caused the change in colour as a visible manifestation of internal thoughts, rendered external for purposes of communication. The fancy I am adding is that the squid's thoughts themselves may reside nowhere but in the skin. If squids think with their skins they are even more 'Martian' than my colleague realized. Even if that is too far-fetched a speculation (it is), the spectacle of their rippling colour changes is quite alien enough to jolt us out of our anaesthetic of familiarity.

Squids are not the only 'Martians' on our own doorstep. Think of the grotesque faces of deep-sea fish, think of dust mites, even more fearsome were they not so tiny; think of basking sharks, just fearsome. Think, indeed, of chameleons with their catapult-launched tongues, swivelling eye turrets and cold, slow gait. Or we can capture that 'strange other world' feeling just as effectively by looking inside ourselves, at the cells that make up our own bodies. A cell is not just a bag of juice. It is packed with solid structures, mazes of intricately folded membranes. There are about 100 million

million cells in a human body, and the total area of membranous structure inside one of us works out at more than 200 acres. That's a respectable farm.

What are all these membranes doing? They seem to stuff the cell as wadding, but that isn't all they do. Much of the folded acreage is given over to chemical production lines, with moving conveyor belts, hundreds of stages in cascade, each leading to the next in precisely crafted sequences, the whole driven by fast-turning chemical cogwheels. The Krebs cycle, the 9-toothed cogwheel that is largely responsible for making energy available to us, turns over at up to 100 revolutions per second, duplicated thousands of times in every cell. Chemical cogwheels of this particular marque are housed inside mitochondria, tiny bodies that reproduce independently inside our cells like bacteria. As we shall see, it is now widely accepted that the mitochondria, along with other vitally necessary structures within cells, not only resemble bacteria but are directly descended from ancestral bacteria who, a billion years ago, gave up their freedom. Each one of us is a city of cells, and each cell a town of bacteria. You are a gigantic megalopolis of bacteria. Doesn't that lift the anaesthetic's pall?

As a microscope helps our minds to burrow through alien galleries of cell membranes, and as a telescope lifts us to far galaxies, another way of coming out of the anaesthetic is to return, in our imaginations, through geological time. It is the inhuman age of fossils that knocks us back on our heels. We pick up a trilobite and the books tell us it is 500 million years old. But we fail to comprehend such an age, and there is a yearning pleasure in the attempt. Our brains have evolved to grasp the time-scales of our own lifetimes. Seconds, minutes, hours, days and years are easy for us. We can cope with centuries. When we come to millennia – thousands of years – our spines begin to tingle. Epic myths of Homer; deeds of the Greek gods Zeus, Apollo and Artemis; of the Jewish heroes Abraham, Moses and David, and their terrifying god Yahweh; of the ancient Egyptians and the Sun God Ra: these inspire poets and give us that *frisson* of immense age. We seem to be peering back through eerie mists into the echoing strangeness of antiquity. Yet, on the time-scale of our trilobite, those vaunted antiquities are scarcely yesterday.

Many dramatizations have been offered, and I shall essay another. Let us write the history of one year on a single sheet of paper. That doesn't leave much room for detail. It is roughly equivalent to the lightning 'Round-up of the Year' that newspapers trot out on 31 December. Each month gets a few sentences. Now on another sheet of paper write the history of the previous year. Carry on back through the years, sketching, at a rate of a year per sheet, the outline of what happened in each year. Bind the pages into a book and number them. Gibbon's *Decline and Fall of the Roman Empire* (1776–88) spans some 13 centuries in six volumes of about 500 pages each, so it is covering the ground at approximately the rate we are talking about.

'Another damned, thick, square book. Always scribble, scribble, scribble! Eh! Mr Gibbon?' WILLIAM HENRY,
FIRST DUKE OF GLOUCESTER (1829)

That splendid volume *The Oxford Dictionary of Quotations* (1992), from which I have just copied this remark, is itself a damned thick, square doorstop of a book, and about the right size to take us back to the time of Queen Elizabeth I. We have an approximate yardstick of time: 4 inches or 10 cm of book thickness to record the history of one millennium. Having established our yardstick, let's work back to the alien world of geological deep time. We place the book of the most recent past flat on the ground, then stack books of earlier centuries on top of it. We now stand beside the pile of books as a living yardstick. If we want to read about Jesus, say, we must select a volume 20 cm from the ground or just above the ankle.

A famous archaeologist dug up a bronze-age warrior with a beautifully preserved face mask and exulted: 'I have gazed upon the face of Agamemnon.' He was being poetically awed at his penetration of fabled antiquity. To find Agamemnon in our pile of books, you'd have to stoop to a level about halfway up your shins. Somewhere in the vicinity you'd find Petra ('A rose-red city, half as old as time'), Ozymandias, king of kings ('Look on my works, ye Mighty, and despair') and that enigmatic wonder of the ancient world the Hanging Gardens of Babylon. Ur of the Chaldees, and Uruk the city of the legendary hero Gilgamesh had their day slightly earlier and you'd find tales of their foundation a little higher up your legs. Around here is the oldest date of all, according to the seventeenth-century archbishop James Ussher, who calculated 4004 BC as the date of the creation of Adam and Eve.

The taming of fire was climacteric in our history; from it stems most of technology. How high in our stack of books is the page on which this epic discovery is recorded? The answer is quite a surprise when you recall that you could comfortably sit down on the pile of books encompassing the whole of recorded history. Archaeological traces suggest that fire was discovered by our *Homo erectus* ancestors, though whether they made fire, or just carried it about and used it we don't know. They had fire by half a million years ago so, to consult the volume in our analogy recording the discovery you'd have to climb up to a level somewhat higher than the Statue of Liberty. A dizzy height, especially given that Prometheus, the legendary bringer of fire, gets his first mention a little below your knee in our pile of books. To read about Lucy and our australopithecine ancestors in Africa, you'd need to climb higher than any building in Chicago. The biography of the common ancestor we share with chimpanzees would be a sentence in a book stacked twice as high again.

But we've only just begun our journey back to the trilobite. How high would the stack of books have to be in order to accommodate the page where the life and death of this trilobite, in its shallow Cambrian sea, is perfunctorily celebrated? The answer is about 56 kilometres, or 35 miles. We aren't used to dealing with heights like this. The summit of mount Everest is less than 9 km above sea level. We can get some idea of the age of the trilobite if we topple the stack through 90 degrees. Picture a bookshelf three times the length of Manhattan island, packed with volumes the size of Gibbon's *Decline and Fall.* To read your way back to the trilobite, with only one page allotted to each year, would be more laborious than spelling through all 14 million volumes in the Library of Congress. But even the trilobite is young compared with the age of life itself. The first living creatures, the shared ancestors of the trilobite, of bacteria and of ourselves, have their ancient chemical lives recorded in volume 1 of our saga. Volume 1 is at the far end of the marathon bookshelf. The entire shelf would stretch from London to the Scottish borders. Or right across Greece from the Adriatic to the Aegean.

Perhaps these distances are still unreal. The art in thinking of analogies for large numbers is not to go off the scale of what people can comprehend. If we do that, we are no better off with an analogy than with the real thing. Reading your way through a work of history, whose shelved volumes stretch from Rome to Venice, is an incomprehensible task, just about as incomprehensible as the bald figure 4,000 million years.

Here is another analogy, one that has been used before. Fling your arms wide in an expansive gesture to span all of evolution from its origin at your left fingertip to today at your right fingertip. All the way across your midline to well past your right shoulder, life consists of nothing but bacteria. Many-celled, invertebrate life flowers somewhere around your right elbow. The dinosaurs originate in the middle of your right palm, and go extinct around your last finger joint. The whole story of *Homo sapiens* and our predecessor *Homo erectus* is contained in the thickness of one nail-clipping. As for recorded history; as for the Sumerians, the Babylonians, the Jewish patriarchs, the dynasties of Pharaohs, the legions of Rome, the Christian Fathers, the Laws of the Medes and Persians which never change; as for Troy and the Greeks, Helen and Achilles and Agamemnon dead; as for Napoleon and Hitler, the Beatles and Bill Clinton, they and everyone that knew them are blown away in the dust from one light stroke of a nail-file.

> *The poor are fast forgotten,*
> *They outnumber the living, but where are all their bones?*
> *For every man alive there are a million dead,*
> *Has their dust gone into earth that it is never seen?*
> *There should be no air to breathe, with it so thick,*
> *No space for wind to blow, nor rain to fall;*
> *Earth should be a cloud of dust, a soil of bones,*
> *With no room even, for our skeletons.*
>
> SACHEVERELL SITWELL, 'Agamemnon's Tomb' (1933)

Not that it matters, Sitwell's third line is inaccurate. It has been estimated that the people alive today make up a substantial proportion of the humans that have ever lived. But this just reflects the power of exponential growth. If we count generations instead of bodies, and especially if we go back beyond humankind to life's beginning, Sacheverell Sitwell's sentiment has a new force. Let us suppose that each individual in our direct female ancestry, from the first flowering of many-celled life a little over half a billion years ago, lay down and died on the grave of her mother, eventually to be fossilized. As in the successive layers of the buried city of Troy, there would be much compression and shaking down, so let us assume that each fossil in the series was flattened to the thickness of a 1 cm pancake. What depth of rock should we need, if we are to accommodate our continuous fossil record? The answer is that the rock would have to be about 1,000 km or 600 miles thick. This is about ten times the thickness of the earth's crust.

The Grand Canyon, whose rocks, from deepest to shallowest, span most of the period we are now talking about, is only around one mile deep. If the strata of the Grand Canyon were stuffed with fossils and no intervening rock, there would be room within its depth to accommodate only about one 600th of the generations that have successively died. This calculation helps us to keep in proportion fundamentalist demands for a 'continuous' series of gradually changing fossils before they will accept the fact of evolution. The rocks of the earth simply don't have room for such a luxury – not by many orders of magnitude. Whichever way you look at it, only an extremely small proportion of creatures has the good fortune to be fossilized. As I have said before, I should consider it an honour.

The number of the dead long exceedeth all that shall live. The night of time far surpasseth the day, and who knows when was the Aequinox? Every houre addes unto that current Arithmetique, which scarce stands one moment . . . Who knows whether the best of men be known, or whether there be not more remarkable persons forgot than any that stand remembred in the known account of time?

SIR THOMAS BROWNE, *Urne Buriall* (1658)

2

DRAWING ROOM OF DUKES

You may grind their souls in the self-same mill,
You may bind them, heart and brow;
But the poet will follow the rainbow still,
And his brother will follow the plow.

JOHN BOYLE O'REILLY (1844–90)
'The Rainbow's Treasure'

Breaking through the anaesthetic of familiarity is what poets do best. It is their business. But poets, too many of them and for too long, have overlooked the goldmine of inspiration offered by science. W. H. Auden, leader of his generation of poets, was flatteringly sympathetic to scientists but even he singled out their practical side, comparing scientists, to their advantage, with politicians, but missing the poetic possibilities of science itself.

The true men of action in our time, those who transform the world, are not the politicians and statesmen, but the scientists. Unfortunately poetry cannot celebrate them, because their deeds are concerned with things, not persons, and are, therefore, speechless. When I find myself in the company of scientists, I feel like a shabby curate who has strayed by mistake into a drawing room full of dukes.

The Dyer's Hand, 'Poet and the City' (1963)

Ironically that is pretty much how I and many other scientists feel when in the company of poets. Indeed – and I shall return to the point – this is probably our culture's normal evaluation of the relative standings of scientists and poets, which may have been why Auden bothered to say the opposite. But why was he so definite that poetry cannot celebrate scientists and their deeds? Scientists may transform the world more effectively than politicians and statesmen, but that is not all they do, and certainly not all they could do. Scientists transform the way we think about the larger universe. They assist the imagination back to the hot birth of time and forward to the eternal cold, or, in Keats's words, to 'spring direct towards the galaxy'. Isn't the speechless universe a worthy theme? Why would a poet celebrate only persons, and not the slow grind of natural forces that made them? Darwin tried manfully, but Darwin's talents lay elsewhere than in poetry:

It is interesting to contemplate an entangled bank, clothed with many plants of many kinds, with birds singing on the bushes, with various insects flitting about, and with worms crawling through the damp earth, and reflect that these elaborately constructed forms, so different from each other, and dependent on each other in so complex a manner, have all been produced by laws acting around us . . . Thus, from the war of nature, from famine and death, the most exalted object which we are capable of conceiving, namely, the production of the higher animals, directly follows. There is grandeur in this view of life, with its several powers, having been originally breathed into a few forms or into one; and that, whilst this planet has gone cycling on according to the fixed law of gravity, from so simple a beginning endless forms most beautiful and most wonderful have been, and are being, evolved.

On the Origin of Species (1859)

William Blake's interests were religious and mystical but, word for word, I wish I had written the following famous quatrain and, if I had, my inspiration and meaning would have been very different.

To see a world in a grain of sand
And a heaven in a wild flower
Hold infinity in the palm of your hand
And eternity in an hour.

'Auguries of Innocence' (*c.* 1803)

The stanza can be read as all about science, all about standing in the moving spotlight, about taming space and time, about the very large built from the quantum graininess of the very small, a lone flower as a miniature of all evolution. The impulses to awe, reverence and wonder which led Blake to mysticism (and lesser figures to paranormal superstition, as we shall see) are precisely those that lead others of us to science. Our interpretation is different but what excites us is the same. The mystic is content to bask in the wonder and revel in a mystery that we were not 'meant' to understand. The scientist feels the same wonder but is restless, not content; recognizes the mystery as profound, then adds, 'But we're working on it.'

Blake did not love science, even feared and despised it:

For Bacon and Newton, sheath'd in dismal steel, their terrors hang
Like iron scourges over Albion; Reasonings like vast Serpents
Infold around my limbs . . .

'Bacon, Newton, and Locke', *Jerusalem* (1804–20)

What a waste of poetic talent. And if, as fashionable commentators can be relied upon to insist, a political motive underlay his poem, it is still a waste; for politics and its preoccupations are so temporary, so trifling by comparison. It is my thesis that poets could better use the inspiration provided by science and that at the same time scientists must reach out to the constituency that I am identifying with, for want of a better word, poets.

It is not, of course, that science should be declaimed in verse.

The rhyming couplets of Erasmus Darwin, Charles's grandfather, though surprisingly well regarded in their time, do not enhance the science. Nor, unless scientists happen to have the talents of a Carl Sagan, a Peter Atkins or a Loren Eiseley, should they cultivate a deliberately prose-poetic style in their expositions. Simple, sober clarity will do nicely, letting the facts and the ideas speak for themselves. The poetry is in the science.

Poets can be obscure, sometimes for good reason, and they rightly claim immunity from the obligation to explain their lines. 'Tell me Mr Eliot, how exactly does one measure out one's life with coffee spoons?' would not, to say the least, have been a good conversation opener, but a scientist, rightly, expects to be asked equivalent questions. 'In what sense can a gene be selfish?' 'What exactly flows down the River Out of Eden?' I still spell out on demand the meaning of Mount Improbable and how slowly and gradually it is climbed. Our language must strive to enlighten and explain, and if we fail to convey our meaning by one approach we should go to work on another. But, without losing lucidity, indeed with added lucidity, we need to reclaim for real science that style of awed wonder that moved mystics like Blake. Real science has a just entitlement to the tingle in the spine which, at a lower level, attracts the fans of *Star Trek* and *Doctor Who* and which, at the lowest level of all, has been lucratively hijacked by astrologers, clairvoyants and television psychics.

Hijacking by pseudo-scientists is not the only threat to our sense of wonder. Populist 'dumbing down' is another, and I shall return to it. A third is hostility from academics sophisticated in fashionable disciplines. A voguish fad sees science as only one of many cultural myths, no more true nor valid than the myths of any other culture. In the United States it is fed by justified guilt over the historical treatment of Native Americans. But the consequences can be laughable; as in the case of Kennewick Man.

Kennewick Man is a skeleton discovered in Washington State in 1996, carbon-dated to older than 9,000 years. Anthropologists were intrigued by anatomical suggestions that he might be unrelated to typical Native Americans, and therefore might represent a separate early migration across what is now the Bering Strait, or even from Iceland. They were preparing to do all-important DNA tests when the legal authorities seized the skeleton, intending to hand it over to representatives of local Indian tribes, who proposed to bury it and forbid all further study. Naturally there was widespread opposition from the scientific and archaeological community. Even if Kennewick Man is an American Indian of some kind, it is highly unlikely that his affinities lie with whichever particular tribe happens to live in the same area 9,000 years later.

Native Americans have impressive legal muscle, and 'The Ancient One' might have been handed over to the tribes, but for a bizarre twist. The Asatru Folk Assembly, a group of worshippers of the Norse gods Thor and Odin, filed an independent legal claim that Kennewick Man was actually a Viking. This Nordic sect, whose views you may follow in the Summer 1997 issue of *The Runestone*, were actually allowed to hold a religious service over the bones. This upset the Yakama Indian community, whose spokesman feared that the Viking ceremony could be 'keeping Kennewick Man's spirit from finding his body'. The dispute between Indians and Norsemen could well be settled by DNA comparison, and the Norsemen are quite keen to be put to this test. Scientific study of the remains would certainly cast fascinating light on the question of when humans first arrived in America. But Indian leaders resent the very idea of studying this question, because they believe their ancestors have been in America since the creation. As Armand Minthorn, religious leader of the Umatilla tribe, put it: 'From our oral histories, we know that our people have been part of this land since the beginning of time. We do not believe our people migrated here from another continent, as the scientists do.'

Perhaps the best policy for the archaeologists would be to declare themselves a religion, with DNA fingerprints their sacramental totem. Facetious but, such is the climate in the United States at the end of the twentieth century, it is possibly the only recourse that would work. If you say, 'Look, here is overwhelming evidence from carbon dating, from mitochondrial DNA, and from archaeological analyses of pottery, that X is the case' you will get nowhere. But if you say, 'It is a fundamental and unquestioned belief of my culture that X is the case' you will immediately hold a judge's attention.

It will also hold the attention of many in the academic community who, in the late twentieth century, have discovered a new form of anti-scientific rhetoric, sometimes called the 'post-modern critique' of science. The most thorough whistle-blowing on this kind of thing is Paul Gross and Norman Levitt's splendid book *Higher Superstition: The Academic Left and its Quarrels with Science* (1994). The American anthropologist Matt Cartmill sums up the basic credo:

> *Anybody who claims to have objective knowledge about anything is trying to control and dominate the rest of us . . . There are no objective facts. All supposed 'facts' are contaminated with theories, and all theories are infested with moral and political doctrines . . . Therefore, when some guy in a lab coat tells you that such and such is an objective fact . . . he must have a political agenda up his starched white sleeve.*
>
> 'Oppressed by evolution', *Discover* magazine (1998)

There are even a few vocal fifth columnists within science itself who hold exactly these views, and use them to waste the time of the rest of us.

Cartmill's thesis is that there is an unexpected and pernicious alliance between the know-nothing fundamentalist religious right and the sophisticated academic left. A bizarre manifestation of the alliance is their joint opposition to the theory of evolution. The opposition of the fundamentalists is obvious. That of the left is a compound of hostility to science in general, of 'respect' (weasel word of our time) for tribal creation myths, and of various political agendas. Both these strange bedfellows share a concern for 'human dignity' and take offence at treating humans as 'animals'. Barbara Ehrenreich and Janet McIntosh make a similar point about what they call 'secular creationists' in their 1997 article 'The New Creationism' in *The Nation* magazine.

Purveyors of cultural relativism and the 'higher superstition' are apt to pour scorn on the search for truth. This partly stems from the conviction that truths are different in different cultures (that was the point of the Kennewick Man story) and partly from the inability of philosophers of science to agree about truth anyway. There are, of course, genuine philosophical difficulties. Is a truth just a so-far-unfalsified hypothesis? What status does truth have in the strange, uncertain world of quantum theory? Is anything ultimately true? On the other hand, no philosopher has any trouble using the language of truth when falsely accused of a crime, or when suspecting his wife of adultery. 'Is it true?' feels like a fair question, and few who ask it in their private lives would be satisfied with logic-chopping sophistry in response. Quantum thought experimenters may not know in what sense it is 'true' that Schrödinger's cat is dead. But everybody knows what is true about the statement that my childhood cat Jane is dead. And there are lots of scientific truths where what we claim is only that they are true in the same everyday sense. If I tell you that humans and chimpanzees share a common ancestor, you may doubt the truth of my statement and search (in vain) for evidence that it is false. But we both know what it would mean for it to be true, and what it would mean for it to be false. It is in the same category as 'Is it true that you were in Oxford on the night of the crime?', not in the same difficult category as 'Is it true that a quantum has position?' Yes, there are philosophical difficulties about truth, but we can get a long way before we have to worry about them. Premature erection of alleged philosophical problems is sometimes a smokescreen for mischief.

'Dumbing down' is a very different kind of threat to scientific sensibility. The 'Public Understanding of Science' movement, provoked in America by the Soviet Union's triumphant entry into the space race and driven today, at least in Britain, by public alarm over a decline in applications for science places at universities, is going demotic. 'Science Weeks' and 'Science Fortnights' betray an anxiety among scientists to be loved. Funny hats and larky voices proclaim that science is fun, fun, fun. Whacky 'personalities' perform explosions and funky tricks. I recently attended a briefing session where scientists were urged to put on events in shopping malls designed to lure people into the joys of science. The speaker advised us to do nothing that might conceivably be seen as a turn-off. Always make your science 'relevant' to ordinary people's lives, to what goes on in their own kitchen or bathroom. Where possible, choose experimental materials that your audience can eat at the end. At the last event organized by the speaker himself, the scientific phenomenon that really grabbed attention was the urinal that automatically flushed as you stepped away. The very word science is best avoided, we were told, because 'ordinary people' see it as threatening.

I have little doubt that such dumbing down will be successful if our aim is to maximize the total population count at our 'event'. But when I protest that what is being marketed here is not real science, I am rebuked for my 'elitism' and told that luring people in, by any means, is a necessary first step. Well, if we must use the word (I wouldn't), maybe elitism is not such a terrible thing. And there is a great difference between an exclusive snobbery and an embracing, flattering elitism that strives to help people to raise their game and join the elite. A calculated dumbing down is the worst: condescending and patronizing. When I gave these views in a recent lecture in America, a questioner at the end, no doubt with a glow of political self-congratulation in his white male heart, had the insulting impertinence to suggest that dumbing down might be necessary to bring 'minorities and women' to science.

I worry that to promote science as all fun and larky and easy is to store up trouble for the future. Real science can be hard (well, challenging, to give it a more positive spin) but, like classical literature or playing the violin, worth the struggle. If children are lured into science, or any other worthwhile occupation, by the promise of easy fun, what are they going to do when they finally have to confront the reality? Recruiting advertisements for the army rightly don't promise a picnic: they seek young people dedicated enough to stand the pace. 'Fun' sends the wrong signals and might attract people to science for the wrong reasons. Literary scholarship is in danger of becoming similarly undermined. Idle students are seduced into a debased 'Cultural Studies', on the promise that they will spend their time deconstructing soap operas, tabloid princesses and Tellytubbies. Science, like proper literary studies, can be hard and challenging but science is – also like proper literary studies – wonderful. Science *can* pay its way but, like great art, it shouldn't have to. And we shouldn't need whacky personalities and fun explosions to persuade us of the value of a life spent finding out why we have life in the first place.

I fear that I may have been too negative in this attack, but there are times when a pendulum has swung far enough and needs a strong push in the other direction to restore equilibrium. *Of course* science is fun, in the sense that it is the very opposite of boring. It can enthral a good mind for a lifetime. *Certainly*, practical demonstrations can help to make ideas vivid and lasting in the mind. From Michael Faraday's Royal Institution Christmas Lectures to Richard Gregory's Bristol Exploratory, children have been excited by hands-on experience of true science. I have myself been honoured to give the Christmas Lectures, in their modern televised form, and I depended upon plenty of hands-on demonstrations. Faraday never dumbed down. I am attacking only the kind of populist whoring that defiles the wonder of science.

Annually there is a large dinner in London at which prizes for the year's best popular science books are awarded. One prize is for children's books on science, and it was recently won by a book about insects and other 'horrible ugly bugs'. That kind of language is perhaps not best calculated to arouse the poetic sense of wonder, but let us be tolerant and acknowledge other ways of attracting the interest of children. Harder to forgive were the antics of the chairman of the judges, a well-known television personality (who had recently sold out to the lucrative genre of 'paranormal' television). Squeaking with game-show levity, she incited the large audience (of adults) to join her in repeated choruses of audible grimaces at the contemplation of the horrible 'ugly bugs'. 'Eeeu-urrrgh! Yuck! Yeeyuck! Eeeeeuurrrgh!' That kind of vulgar fun demeans the wonder of science, and risks 'turning off' the very people best qualified to appreciate it and inspire others: real poets and true scholars of literature.

By poets, of course, I intend artists of all kinds. Michelangelo and Bach were paid to celebrate the sacred themes of their times and the results will always strike human senses as sublime. But we shall never know how such genius might have responded to alternative commissions. As Michelangelo's mind moved upon silence 'Like a long-legged fly upon the stream', what might he not have painted if he had known the contents of one nerve cell from a long-legged fly? Think of the 'Dies Irae' that might have been wrung from Verdi by the contemplation of the dinosaurs' fate when, 65 million years ago, a mountain-sized rock screamed out of deep space at 10,000 miles per hour straight at the Yucatán peninsula and the world went dark. Try to imagine Beethoven's 'Evolution Symphony', Haydn's oratorio on 'The Expanding Universe', or Milton's epic *The Milky Way*. As for Shakespeare . . . But we don't have to aim so high. Lesser poets would be a fine start.

I can imagine, in some otherworld
Primeval-dumb, far back
In that most awful stillness, that only gasped and hummed,
Humming-birds raced down the avenues.

Before anything had a soul,
While life was a heave of matter, half inanimate,
This little bit chipped off in brilliance
And went whizzing through the slow, vast, succulent stems.

I believe there were no flowers then,
In the world where the humming-bird flashed ahead of creation.
I believe he pierced the slow vegetable veins with his long beak.

Probably he was big
As mosses, and little lizards, they say, were once big.
Probably he was a jabbing, terrifying monster.

We look at him through the wrong end of the telescope of Time,
Luckily for us. *Unrhyming Poems*, 1928

D. H. Lawrence's poem about hummingbirds is almost wholly inaccurate and therefore, superficially, unscientific. Yet, in spite of this, it is a passable shot at how a poet might take inspiration from geological time. Lawrence lacked only a couple of tutorials in evolution and taxonomy to bring his poem within the pale of accuracy, and it would be no less arresting and thought-provoking as a poem. After another tutorial Lawrence, the miner's son, might have turned fresh eyes on his coal fire, whose glowing energy last saw the light of day – *was* the light of day – when it warmed the Carboniferous treeferns, to be laid down in earth's dark cellar and sealed for three million centuries. A larger obstacle would have been Lawrence's hostility to what he wrongly thought of as the anti-poetic spirit of science and scientists, as when he grumbled that

Knowledge has killed the sun, making it a ball of gas with spots . . . The world of reason and science . . . this is the dry and sterile world the abstracted mind inhabits.

I am almost reluctant to admit that my favourite of all poets is that confused Irish mystic William Butler Yeats. In old age Yeats sought a theme and sought for it in vain, finally returning, in desperation, to enumerate old themes of his *fin de siècle* young manhood. How sad to give up, wrecked among heathen dreams, marooned amid the faeries and fey Irishry of his affected youth when, an hour's drive from Yeats's tower, Ireland housed the largest astronomical telescope then built. This was the 72-inch reflector, built before Yeats was born by William Parsons, third earl of Rosse, at Birr Castle (where it has now been restored by the seventh earl). What might a single glance at the Milky Way through the eyepiece of the 'Leviathan of Parsonstown' not have done for the frustrated poet who, as a young man, had written these unforgettable lines?

> *Be you still, be you still, trembling heart;*
> *Remember the wisdom out of the old days:*
> *Him who trembles before the flame and the flood,*
> *And the winds that blow through the starry ways,*
> *Let the starry winds and the flame and the flood*
> *Cover over and hide, for he has no part*
> *With the lonely, majestical multitude.*
>
> from *The Wind Among the Reeds* (1899)

Those would make fine last words for a scientist, as would, now that I think about it, the poet's own epitaph, 'Cast a cold eye/On life, on death./ Horseman, pass by!' But, like Blake, Yeats was no lover of science, dismissing it (absurdly), as the 'opium of the suburbs', and calling us to 'Move upon Newton's town.' That is sad, and the kind of thing that drives me to write my books.

Keats, too, complained that Newton had destroyed the poetry of the rainbow by explaining it. By more general implication, science is poetry's killjoy, dry and cold, cheerless, overbearing and lacking in everything that a young Romantic might desire. To proclaim the opposite is one purpose of this book, and I shall here limit myself to the untestable speculation that Keats, like Yeats, might have been an even better poet if he had gone to science for some of his inspiration.

It has been pointed out that Keats's medical education may have equipped him to recognize the mortal symptoms of his own tuberculosis, as when he ominously diagnosed his own arterial blood. Science, for him, would not have been the bringer of good news, so it is less wonder if he found solace in an antiseptic world of classical myth, losing himself among panpipes and naiads, nymphs and dryads, just as Yeats was to do among their Celtic counterparts. Irresistible as I find both poets, forgive my wondering whether the Greeks would have recognized their legends in Keats, or the Celts theirs in Yeats. Were these great poets as well served as they could have been by their sources of inspiration? Did prejudice against reason weigh down the wings of poesy?

It is my thesis that the spirit of wonder which led Blake to Christian mysticism, Keats to Arcadian myth and Yeats to Fenians and fairies, is the very same spirit that moves great scientists; a spirit which, if fed back to poets in scientific guise, might inspire still greater poetry. In support, I adduce the less elevated genre of science fiction. Jules Verne, H. G. Wells, Olaf Stapledon, Robert Heinlein, Isaac Asimov, Arthur C. Clarke, Ray Bradbury and others have used prose-poetry to evoke the romance of scientific themes, in some cases explicitly linking them to the myths of antiquity. The best of science fiction seems to me an important literary form in its own right, snobbishly underrated by some scholars of literature. More than one reputable scientist has been introduced to what I am calling the spirit of wonder through an early fascination with science fiction.

At the lower end of the science fiction market the same spirit has been abused for more sinister ends, but the bridge to mystical and romantic poetry can still be discerned. At least one major religion, Scientology, was founded by a science fiction writer, L. Ron Hubbard (whose entry in the *Oxford Dictionary of Quotations* reads, 'If you really want to make a million . . . the quickest way is to start your own religion'). The now dead adherents of the cult of 'Heaven's Gate' probably never knew that the phrase appears twice in Shakespeare and twice in Keats, but they knew all about *Star Trek* and were obsessed with it. The language of their website is a preposterous caricature of misunderstood science, laced with bad romantic poetry.

The cult of *The X-Files* has been defended as harmless because it is, after all, only fiction. On the face of it, that is a fair defence. But regularly recurring fiction – soap operas, cop series and the like – are legitimately criticized if, week after week, they systematically present a one-sided view of the world. *The X-Files* is a television series in which, every week, two FBI agents face a mystery. One of the two, Scully, favours a rational, scientific explanation; the other agent, Mulder, goes for an explanation which is either supernatural or, at very least, glorifies the inexplicable. The problem with *The X-Files* is that routinely, relentlessly, the supernatural explanation, or at least the Mulder end of the spectrum, usually turns out to be the answer. I'm told that, in recent episodes, even the sceptical agent Scully is starting to have her confidence shaken, and no wonder.

But isn't it just harmless fiction, then? No, I think the defence rings hollow. Imagine a television series in which two police officers solve a crime each week. Every week there is one black suspect and one white suspect. One of the two detectives is always biased towards the black suspect, the other biased towards the white. And, week after week, the black suspect turns out to have done it. So, what's wrong with that? After all, it's only fiction! Shocking as it is, I believe the analogy to be a completely fair one. I am not saying that supernaturalist propaganda is as dangerous or unpleasant as racist propaganda. But *The X-Files* systematically purveys an anti-rational view of the world which, by virtue of its recurrent persistence, is insidious.

Another bastard form of science fiction converges upon Tolkienian faked-up myth. Physicists rub shoulders with wizards, interplanetary aliens escort princesses sidesaddle on unicorns, thousand-port-holed space stations loom out of the same mist as medieval castles with ravens (or even pterodactyls) wheeling around their gothic turrets. True, or calculatedly modified, science is replaced by magic, which is the easy way out.

Good science fiction has no dealings with fairy-tale magic spells, but is premised on the world as an orderly place. There is mystery, but the universe is not frivolous nor light-fingered in its changeability. If you put a brick on a table it stays there unless something moves it, even if you have forgotten it is there. Poltergeists and sprites don't intervene and hurl it about for reasons of mischief or caprice. Science fiction may tinker with the laws of nature, advisedly and preferably one law at a time, but it cannot abolish lawfulness itself and remain good science fiction. Fictional computers may become consciously malevolent or even, in Douglas Adams's masterly science comedies, paranoid; spaceships may warp-drive themselves to distant galaxies using some postulated future technology, but the decencies of science are essentially maintained. Science allows mystery but not magic, strangeness beyond wild imagining but no spells or witchery, no cheap and easy miracles. Bad science fiction loses its grip on moderated lawfulness and substitutes the 'anything goes' profligacy of magic. The worst of bad science fiction joins hands with the 'paranormal', that other lazy, misbegotten child of the sense of wonder which ought to be motivating true science. The popularity of this kind of pseudo-science at least seems to suggest that the sense of wonder is widespread and heartfelt, however misapplied it may be. Here lies the only consolation I can find in the pre-millennial media obsession with the paranormal; with the immensely successful *X-Files* and with popular television shows in which routine conjuring tricks are misrepresented as violating natural law.

But let us return to Auden's pleasing compliment and our inversion of it. Why do some scientists feel like shabby curates among literary dukes, and why do many in our society perceive them so?

Undergraduates specializing in science at my own university have occasionally remarked to me (wistfully, for peer pressure in their cohort is strong) that their subject is not seen as 'cool'. This was illustrated for me by a smart young journalist whom I met on a recent BBC television discussion series. She seemed almost intrigued to meet a scientist, for she confided that when at Oxford she had never known any. Her circle had regarded them from a distance as 'grey men', especially pitying their habit of getting out of bed before lunch. Of all absurd excesses, they attended 9 a.m. lectures and then worked through the morning in the labs. That great humanist and humanitarian statesman Jawaharlal Nehru, as befits the first prime minister of a country that cannot afford to mess about, had a more realistic view of science.

> *It is science alone that can solve the problems of hunger and poverty, of insanitation and illiteracy, of superstition and deadening custom and tradition, of vast resources running to waste, or a rich country inhabited by starving people ... Who indeed could afford to ignore science today? At every turn we have to seek its aid ... The future belongs to science and those who make friends with science**.
>
> (1962)

Nevertheless, the confidence with which scientists sometimes state how much we know and how useful science can be, may spill over into arrogance. The distinguished embryologist Lewis Wolpert once admitted that science is occasionally arrogant, and he went on to remark, mildly, that science has a certain amount to be arrogant about. Peter Medawar, Carl Sagan and Peter Atkins have all said something similar.

* Correcting copy in August 1998, I cannot let this pass without sadly reflecting that Nehru would feel India's decision to carry out nuclear tests, unilaterally and in defiance of world opinion, to be a shocking abuse of science and a desecration of his memory and that of Mahatma Gandhi.

Arrogant or not, we at least pay lip-service to the idea that science advances by *disproof* of its hypotheses. Konrad Lorenz, father of ethology, characteristically exaggerated when he said he looked forward to disproving at least one pet hypothesis daily, before breakfast. But it is true that scientists, more than, say, lawyers, doctors or politicians, gain prestige among their peers by publicly admitting their mistakes. One of the formative experiences of my Oxford undergraduate years occurred when a visiting lecturer from America presented evidence that conclusively disproved the pet theory of a deeply respected elder statesman of our zoology department, the theory that we had all been brought up on. At the end of the lecture, the old man rose, strode to the front of the hall, shook the American warmly by the hand and declared, in ringing emotional tones, 'My dear fellow, I wish to thank you. I have been wrong these fifteen years.' We clapped our hands red. Is any other profession so generous towards its admitted mistakes?

Science progresses by correcting its mistakes, and makes no secret of what it still does not understand. Yet the opposite is widely perceived. Bernard Levin, when a columnist on *The Times* of London, sporadically published tirades against science, and on 11 October 1996 he wrote one headed 'God, me and Dr Dawkins', with the subtitle 'Scientists don't know and nor do I – but at least I know I don't know', above which was a cartoon of me as Michelangelo's Adam encountering the pointing finger of God. But as any scientist would vigorously protest, it is of the essence of science to know what we do not know. This is precisely what drives us to find out. In an earlier column, of 29 July 1994, Bernard Levin had made light of the idea of quarks ('The quarks are coming! The quarks are coming! Run for your lives . . .'). After further cracks about 'noble science' having given us mobile telephones, collapsible umbrellas and multi-striped toothpaste, he broke into mock seriousness:

Can you eat quarks? Can you spread them on your bed when the cold weather comes?

This sort of thing doesn't really deserve a reply, but the Cambridge metallurgist Sir Alan Cottrell gave it two sentences, in a Letter to the Editor a few days later.

Sir: Mr Bernard Levin asks 'Can you eat quarks?' I estimate that he eats 500, 000,000, 000,000, 000,000, 000,001 quarks a day . . . Yours faithfully . . .

Admitting what you don't know is a virtue, but gloating ignorance of the arts on such a scale would, quite rightly, not be tolerated by any editor. Philistine ignorance of science is still, in some quarters, thought witty and clever. How else to explain the following little joke, by a recent editor of the London *Daily Telegraph*? The paper was reporting the dumbfounding fact that a third of the British population still believes that the sun goes round the earth. At this point the editor inserted a note in square brackets: '[Doesn't it? Ed.]' If a survey had shown a third of the British populace believing that Shakespeare wrote *The Iliad*, no editor would humorously feign ignorance of Homer. But it is socially acceptable to boast ignorance of science and proudly claim incompetence in mathematics. I have made the point often enough to sound plaintive, so let me quote Melvyn Bragg, one of the most justly respected commentators on the arts in Britain, from his book about scientists, *On Giants' Shoulders* (1998).

There are still those who are affected enough to say they know nothing about the sciences as if this somehow makes them superior. What it makes them is rather silly, and it puts them at the fag end of that tired old British tradition of intellectual snobbery which considers all knowledge, especially science, as 'trade'.

Sir Peter Medawar, that swashbuckling Nobel Prize-winner whom I've already cited, said something similar about 'trade', vividly lampooning the British distaste for all things practical.

It is said that in ancient China the mandarins allowed their fingernails – or anyhow one of them – to grow so extremely long as manifestly to unfit them for any manual activity, thus making it perfectly clear to all that they were creatures too refined and elevated ever to engage in such employments. It is a gesture that cannot but appeal to the English, who surpass all other nations in snobbishness; our fastidious distaste for the applied sciences and for trade has played a large part in bringing England to the position in the world which she occupies today.

The Limits of Science (1984)

Antipathy to science can become quite pettish. Listen to the novelist and feminist Fay Weldon's hymn of hate against 'the scientists', also in the *Daily Telegraph*, on 2 December 1991. (I imply nothing by this coincidence, for the paper has an energetic science editor and fine coverage of scientific topics):

Don't expect us to like you. You promised us too much and failed to deliver. You never even tried to answer the questions we all asked when we were six. Where did Aunt Maud go when she died? Where was she before she was born?

Note that this accusation is the precise opposite of Bernard Levin's (that scientists don't know when they don't know). If I were to offer a simple and direct best-guess answer to both those Aunt Maud questions, I'd certainly be called arrogant and presumptuous, going beyond what I could possibly know, going beyond the limits of science. Miss Weldon continues:

You think these questions are simplistic and embarrassing, but they're the ones which interest us. And who cares about half a second after the Big Bang; what about half a second before? And what about crop circles? . . . The scientists just can't face the notion of a variable universe. We can.

She never makes clear who this all-inclusive, anti-scientific 'we' is, and she probably, by now, regrets the tone of her piece. But it is worth worrying where such naked hostility comes from.

Another example of anti-science, though in this case possibly intended to be funny, is a piece from A. A. Gill, a humorous loose cannon of a columnist in the *Sunday Times* of London (8 September 1996). He refers to science as constrained by experiment, and by the tedious, plodding stepping stones of empiricism. He contrasts it with art and with the theatre, with the magic of lights, fairy dust, music and applause.

There are stars and there are stars, darling. Some are dull, repetitive squiggles on paper, and some are fabulous, witty, thought-provoking, incredibly popular . . .

'Dull, repetitive squiggles' is a reference to the discovery of pulsars, by Bell and Hewish at Cambridge in 1967. Gill was reviewing a television programme in which the astronomer Jocelyn Bell Burnell recalled the spine-tingling moment when she first knew, looking at the print-out from Anthony Hewish's radio telescope, that she was seeing something hitherto unheard of in the universe. A young woman on the threshold of a career, the 'dull, repetitive squiggles' on her roll of paper spoke to her in tones of revolution. Not something new under the sun: a whole new *kind* of sun, a pulsar. Pulsars spin so fast that, where our planet takes 24 hours to rotate, a pulsar may take a fraction of a second. Yet the beam of energy that brings us the news, sweeping round like a lighthouse with such astonishing speed and clocking the seconds more accurately than a quartz crystal, may take millions of years to reach us. Darling, how too too tedious, how madly *empirical*, my dear! Give me fairy dust at the panto any day.

I do not think that such fretful, shallow antipathy results from the common tendency to shoot the messenger, or to blame science for political misuses like hydrogen bombs. No, the hostility I have quoted sounds to me more personally anguished, almost threatened, beleaguered, fearful of humiliation because science is seen as too difficult to master. Oddly enough, I would not *dare* to go so far as John Carey, professor of English literature at Oxford, when he writes, in the preface to his admirable *Faber Book of Science* (1995):

The annual hordes competing for places on arts courses in British universities, and the trickle of science applicants, testify to the abandonment of science among the young. Though most academics are wary of saying it straight out, the general consensus seems to be that arts courses are popular because they are easier, and that most arts students would simply not be up to the intellectual demands of a science course.

Some of the more mathematical sciences may be hard, but nobody should have trouble understanding the circulation of the blood and the heart's role in pumping it round. Carey relates how he quoted to a class of 30 undergraduates, in their final year reading English at a great university, Donne's lines, 'Knows't thou how blood, which to the heart doth flow,/Doth from one ventricle to the other go?' Carey asked them how, as a matter of fact, the blood *does* flow. None of the 30 could answer, and one tentatively guessed that it might be 'by osmosis'. This is not just wrong. Even more spectacularly, it is dull. Dull compared to the truth that the total length of capillaries round which the heart pumps the blood, from ventricle to ventricle, is more than 50 miles. If 50 miles of tubing are packed inside a human body, you can readily work out how finely and intricately ramified most of those tubes must be. I don't think any true scholar could fail to find this an arresting thought. And unlike, say, quantum theory or relativity, it certainly isn't difficult to understand, though it may be difficult to credit. So I take a more charitable view than Professor Carey and wonder whether these young people had simply been let down by scientists and insufficiently inspired by them. Perhaps an emphasis on practical experiment at school, while excellently suited to some children, may be superfluous or positively counterproductive for those who are equally clever but clever in a different way.

Recently I did a television programme about science in our culture (it was, in fact, the one being reviewed by A. A. Gill). Among the many appreciative letters I received was one which poignantly began: 'I am a clarinet teacher whose only memory of science at school was a long period of studying the bunsen burner.' The letter led me to reflect that it is possible to enjoy the Mozart concerto without being able to play the clarinet. In fact, you can learn to be an expert connoisseur of music without being able to play a note on any instrument. Of course, music would come to a halt if nobody ever learned to play it. But if everybody grew up thinking that music was synonymous with *playing* it, think how relatively impoverished many lives would be.

Couldn't we learn to think of science in the same way? It is certainly important that some people, indeed some of our brightest and best, should learn to do science as a practical subject. But couldn't we also teach science as something to read and rejoice in, like learning how to listen to music rather than slaving over five-finger exercises in order to play it? Keats shied away from the dissecting room, and who can blame him? Darwin did the same. Perhaps if he had been taught in a less practical way, Keats would have been more sympathetic to science and Newton.

It is here that I would seek rapprochement with Britain's best-known journalistic critic of science, Simon Jenkins, former editor of *The Times*. Jenkins is a more formidable adversary than the others I have quoted because he knows what he is talking about. He readily concedes that science books can be inspiring, but he resents the high profile science receives in modern compulsory education syllabuses. In a taped conversation with me in 1996, he said:

I can think of very few science books I've read that I've called useful. What they've been is wonderful. They've actually made me feel that the world around me is a much fuller, much more wonderful, much more awesome place than I ever realized it was. That has been, for me, the wonder of science. That's why science fiction retains its compelling fascination for people. That's why the move of science fiction into biology is so intriguing. I think that science has got a wonderful story to tell. But it isn't useful. It's not useful like a course in business studies or law is useful, or even a course in politics and economics.

Jenkins's view that science is not useful is so idiosyncratic that I shall pass over it. Usually even its sternest critics concede that science is useful, perhaps all too useful, while at the same time missing Jenkins's more important point that it can be wonderful. For them, science in its usefulness undermines our humanity or destroys the mystery on which poetry is sometimes thought to thrive. For another thoughtful British journalist, Bryan Appleyard, writing in 1992, science is doing 'appalling spiritual damage'. It is 'talking us into abandoning ourselves, our true selves'. Which brings me back to Keats and his rainbow, and leads us into the next chapter.

3
BARCODES IN THE STARS

Nor ever yet
The melting rainbow's vernal-tinctur'd hues
To me have shone so pleasing, as when first
The hand of science pointed out the path
In which the sun-beams gleaming from the west
Fall on the wat'ry cloud, whose darksome veil
Involves the orient, and that trickling show'r
Piercing thro' every crystalline convex
Of clust'ring dew-drops to their flight oppos'd,
Recoil at length where concave all behind
Th'internal surface of each glassy orb
Repells their forward passage into air;
That thence direct they seek the radiant goal
From which their course began; and as they strike
In diff'rent lines the gazer's obvious eye,
Assume a diff'rent lustre, thro' the brede
Of colours changing from the splendid rose
To the pale violet's dejected hue.

MARK AKENSIDE,
The Pleasures of Imagination (1744)

In December 1817 the English painter and critic Benjamin Haydon introduced John Keats to William Wordsworth at dinner in his London studio, together with Charles Lamb and others of the English literary circle. On view was Haydon's new painting of Christ entering Jerusalem, attended by the figures of Newton as a believer and Voltaire as a sceptic. Lamb, drunk, upbraided Haydon for painting Newton, 'a fellow who believed nothing unless it was as clear as the three sides of a triangle'. Newton, Keats agreed with Lamb, had destroyed all the poetry of the rainbow, by reducing it to the prismatic colours. 'It was impossible to resist him,' said Haydon, 'and we all drank "Newton's health, and confusion to mathematics".' Years later, Haydon recalled this 'immortal dinner' in a letter to Wordsworth, his fellow survivor.

And don't you remember Keats proposing 'Confusion to the memory of Newton', and upon your insisting on an explanation before you drank it, his saying, 'Because he destroyed the poetry of the rainbow by reducing it to a prism'? Ah, my dear old friend, you and I shall never see such days again!

Haydon,
Autobiography and Memoirs

Three years after Haydon's dinner, in his long poem 'Lamia' (1820), Keats wrote:

Do not all charms fly
At the mere touch of cold philosophy?
There was an awful rainbow once in heaven:
We know her woof, her texture; she is given
In the dull catalogue of common things.
Philosophy will clip an Angel's wings,
Conquer all mysteries by rule and line,
Empty the haunted air, and gnomed mine –
Unweave a rainbow . . .

Wordsworth had better regard for science, and for Newton ('Voyaging through strange seas of thought, alone'). He also, in his preface to the *Lyrical Ballads* (1802), anticipated a time when 'The remotest discoveries of the chemist, the botanist, or mineralogist, will be as proper objects of the poet's art as any upon which it can be employed'.

His collaborator Coleridge said elsewhere that 'the souls of 500 Sir Isaac Newtons would go to the making up of a Shakespeare or a Milton'. This can be interpreted as the naked hostility of a leading Romantic against science in general, but the case of Coleridge is more complicated. He read a great deal of science and fancied himself as a scientific thinker, not least on the subject of light and colour, where he claimed to have anticipated Goethe. Some of Coleridge's scientific speculations have turned out to be plagiarisms, and he perhaps showed poor judgement over whom to plagiarize. It was not scientists in general that Coleridge anathematized, but Newton in particular. He had a high regard for Sir Humphry Davy, whose lectures he attended at the Royal Institution 'in order to renew my stock of metaphors'. He felt that Davy's discoveries, compared with Newton's, were 'more intellectual, more ennobling and impowering human nature'. His use of words like ennobling and impowering suggests that Coleridge's heart might have been in the right place with respect to science, if not with respect to Newton. But he failed to live up to his own ideals 'to unfold and arrange' his ideas in 'distinct, clear and communicable conceptions'. On the subject of the spectrum and unweaving the rainbow itself, in a letter of 1817 he became almost beside himself with confusion:

To me, I confess, Newton's positions, first, of a Ray *of Light, as a physical* synodical Individuum, *secondly, that 7 specific individua are co-existent (by what copula?) in this complex yet divisible Ray; thirdly that the Prism is a mere mechanic Dissector of this Ray; and lastly, that Light, as the common result, is = confusion.*

In another 1817 letter, Coleridge warms to his theme:

So again Colour is Gravitation under the power of Light, Yellow being the positive, blue the negative Pole, and Red the culmination or Equator; while Sound on the other hand is Light under the power or paramountcy of Gravitation.

Perhaps Coleridge was simply born too early to be a post-modernist:

The figure/ground distinction prevalent in Gravity's Rainbow *is also evident in* Vineland, *although in a more self-supporting sense. Thus Derrida uses the term 'subsemioticist cultural theory' to denote the role of the reader as poet. Thus, the subject is contextualized into a postcultural capitalist theory that includes language as a paradox.*

This is from http://www.cs.monash.edu.au/links/postmodern.html where a literally infinite quantity of similar nonsense can be found. The meaningless wordplays of modish francophone *savants*, splendidly exposed in Alan Sokal and Jean Bricmont's *Intellectual Impostures* (1998), seem to have no other function than to impress the gullible. They don't even want to be understood. A colleague confessed to an American devotee of post-modernism that she found his book very difficult to understand. 'Oh, thank you very much,' he smiled, obviously delighted at the compliment. Coleridge's scientific ramblings, by contrast, seem to show some genuine, if incoherent, desire to understand the world around him. We must set him on one side as a unique anomaly, and move on.

Why, in Keats's 'Lamia', is the philosophy of rule and line 'cold', and why do all charms flee before it? What is so threatening about reason? Mysteries do not lose their poetry when solved. Quite the contrary; the solution often turns out more beautiful than the puzzle and, in any case, when you have solved one mystery you uncover others, perhaps to inspire greater poetry. The distinguished theoretical physicist Richard Feynman was charged by a friend that a scientist misses the beauty of a flower by studying it. Feynman responded:

The beauty that is there for you is also available for me, too. But I see a deeper beauty that isn't so readily available to others. I can see the complicated interactions of the flower. The color of the flower is red. Does the fact that the plant has color mean that it evolved to attract insects? This adds a further question. Can insects see color? Do they have an aesthetic sense? And so on. I don't see how studying a flower ever detracts from its beauty. It only adds.

from 'Remembering Richard Feynman', *The Skeptical Inquirer* (1988)

Newton's dissection of the rainbow into light of different wavelengths led on to Maxwell's theory of electromagnetism and thence to Einstein's theory of special relativity. If you think the rainbow has poetic mystery, you should try relativity. Einstein himself openly made aesthetic judgements in science, and perhaps went too far. 'The most beautiful thing we can experience,' he said, 'is the mysterious. It is the source of all true art and science.' Sir Arthur Eddington, whose own scientific writings were noted for poetic flair, used the solar eclipse of 1919 to test General Relativity and returned from Principe Island to announce, in Banesh Hoffmann's phrase, that Germany was host to the greatest scientist of the age. I read those words with a catch in the throat, but Einstein himself took the triumph in his stride. Any other result and he would have been 'sorry for the dear Lord. The theory is correct.'

Isaac Newton made a private rainbow in a dark room. A small hole in a shutter admitted a sunbeam. In its path he placed his famous prism, which refracted (bent) the sunbeam through an angle, once as it penetrated the glass, then again as it passed through the farther facet into the air again. When the light fell on the far wall of Newton's room, the colours of the spectrum were clearly displayed. Newton was not the first to make an artificial rainbow with a prism, but he was the first to use it to demonstrate that white light is a mixture of different colours. The prism sorts them out by bending them through different angles, blue through a steeper angle than red; green, yellow and orange through intermediate angles. Others had, understandably, thought that a prism changed the quality of the light, positively tinting it rather than separating the colours out of an existing mixture. Newton clinched the matter in two experiments in which the light passed through a second prism. In his '*experimentum crucis*', beyond the first prism he placed a slit which allowed only a small part of the spectrum to pass, say, the red portion. When this red light was again refracted by a second prism, only red light emerged. This showed that light is not qualitatively changed by a prism, merely separated out into components which would normally be mixed. In his other clinching experiment, Newton turned the second prism upside down. The spectral colours that had been fanned out by the first prism were brought together again by the second. What emerged was reconstituted white light.

The easiest way to understand the spectrum is through the wave theory of light. The thing about waves is that nothing actually travels all the way from source to destination. Such motion as there is, is local and small scale. Local motion triggers motion in the next local patch and so on, all the way along the line, like the famous football stadium wave. The original wave theory of light was in turn supplanted by the quantum theory, according to which light is delivered as a stream of discrete photons. Physicists that I have pressed admit that photons stream away from the sun in a way that football fans do not travel from one end of the stadium to the other. Nevertheless, ingenious experiments in this century have shown that even in the quantum theory photons still behave like waves too. For many purposes, including ours in this chapter, we can forget quantum theory and treat light simply as waves propagating outwards from a light source, like ripples in a pond when a pebble is thrown in. But light waves travel incomparably faster and are broadcast in three dimensions. To unweave the rainbow is to separate it into its components of different wavelengths. White light is a scrambled mixture of wavelengths, a visual cacophony. White objects reflect light of all wavelengths but, unlike mirrors, they scatter it into incoherence as they do so. This is why you see light, but not your face, reflected from a white wall. Black objects absorb light of all wavelengths. Coloured objects, by reason of the atomic structures of their pigments or surface layers, absorb light of some wavelengths and reflect other wavelengths. Plain glass allows light of all wavelengths to pass straight through it. Coloured glass transmits light of some wavelengths while absorbing light of other wavelengths.

What is it about the bending action of a glass prism or, under the right conditions, a drop of rain, that splits white light into its separate colours? And anyway, why are light beams bent by glass and water at all? The bending results from a slowing down of the light as it moves from air into glass (or water). It speeds up again as it emerges from the glass. How can this be, given Einstein's dictum that the velocity of light is the great physical constant of the universe, and nothing can go faster? The answer is that light's legendary full speed, represented by the symbol *c*, is attained only in a vacuum. When light travels through a transparent substance like glass or water, it is slowed down by a factor known as the 'refractive index' of that substance. It is slowed down by air, too, but less so.

But why does slowing down translate into a change of angle? If the light beam is pointing straight into a glass block, it will continue at the same angle (dead ahead) but slowed down. However, if it breaks the surface at an oblique angle, it is deflected to a shallower angle as it starts to travel more slowly. Why? Physicists have coined a 'Principle of Least Action' which, if not entirely satisfying as an ultimate explanation, at least makes it something we can empathize with. The matter is well explained in Peter Atkins's *Creation Revisited* (1992). Some physical entity, in this case a beam of light, behaves as if it is striving for economy, trying to minimize something. Imagine yourself a lifesaver on a beach, racing to save a drowning child. Every second counts, and you must take as little time as possible to reach the child. You can run faster than you can swim. Your course towards the child is initially over land and therefore fast, then through water and so much slower. Assuming that the child is not straight out to sea from where you are standing, how do you minimize your travel time? You could take the beeline direction, minimizing distance, but this wouldn't minimize the time taken because it leaves too much of the journey in water. You could run straight to that point on the sea's edge which is immediately opposite the child, then swim straight out to sea. This maximizes running at the expense of swimming, but even this is not quite the fastest course because of the greater total distance travelled. It is easy to see that the swiftest course is to run to the shore at a critical angle, which depends upon the ratio of your running speed to your swimming speed, then switch abruptly to a new angle for the swimming part of the journey. In terms of the analogy, swimming speed and running speed correspond to the refractive index of water and the refractive index of air. Of course light beams aren't deliberately 'trying' to minimize their travel time, but everything about their behaviour makes sense if you assume that they are doing the unconscious equivalent. The analogy can be made respectable in terms of quantum theory, but that is beyond my scope here and I recommend Atkins's book.

The spectrum depends upon light of different colours being slowed by different amounts: the refractive index of a given substance, say glass or water, is greater for blue light than for red. You could think of blue light as being a slower swimmer than red, getting tangled up in the undergrowth of atoms in glass or water because of its short wavelength. Light of all colours gets less tangled up among the sparser atoms of air, but blue still travels more slowly than red. In a vacuum, where there is no undergrowth at all, light of all colours has the same velocity: the great, universal maximum *c*.

Raindrops have a more complicated effect than Newton's prism. Being roughly spherical, their back surface acts as a concave mirror. So they reflect the sunlight after refracting it, which is why we see the rainbow in the part of the sky opposite the sun, rather than when looking towards the sun through rain. Imagine that you are standing with your back to the sun, looking towards a shower of rain, preferably with a leaden background. We shan't see a rainbow if the sun is higher in the sky than 42 degrees above the horizon. The lower the sun, the higher the rainbow. As the sun rises in the morning, the rainbow, if one is visible, sets. As the sun sets in the evening, the rainbow rises. So let's assume that it is early morning or late afternoon. Think about a particular raindrop as a sphere. The sun is behind and slightly above you, and light from it enters the raindrop. At the boundary of air with water it is refracted and the different wavelengths that make up the sun's light are bent through different angles, as in Newton's prism. The fanned-out colours go through the interior of the raindrop until they hit its concave far wall, where there they are reflected, back and down. They leave the raindrop again and some of them end up at your eye. As they pass from water back into air they are refracted for a second time, the different colours again being bent through different angles.

So, a complete spectrum – red, orange, yellow, green, blue, violet – leaves our single raindrop, and a similar one leaves the other raindrops in the vicinity. But from any one raindrop, only a small part of the spectrum hits your eye. If your eye gets a beam of green light from one particular raindrop, the blue light from that raindrop goes above your eye, and the red light from that particular raindrop goes below. So, why do you see a complete rainbow? Because there are lots of different raindrops. A band of thousands of raindrops is giving you green light (and simultaneously giving blue light to anybody who might be suitably placed above you, and simultaneously giving red light to somebody else below you). Another band of thousands of raindrops is giving you red light (and giving somebody else blue light . . .), another band of thousands of raindrops is giving you blue light, and so on. The raindrops delivering red light to you are all at a fixed distance from you – which is why the red band is curved (you are the centre of the circle). The raindrops delivering green light to you are also at a fixed distance from you, but it is a shorter one. So the circle on which they sit has a smaller radius and the green curve sits inside the red curve. Then the blue curve sits inside that, and the whole rainbow is built up as a series of circles with you at the centre. Other observers will see different rainbows centred on themselves.

So, far from the rainbow being rooted at a particular 'place' where fairies might deposit a crock of gold, there are as many rainbows as there are eyes looking at the storm. Different observers, looking at the same shower from different places, will piece together their own separate rainbows using light from different collections of raindrops. Strictly speaking, even your two eyes are seeing two different rainbows. And as we drive along a road looking at 'one' rainbow, we are actually seeing a series of rainbows in quick succession. I think that if Wordsworth had realized all this, he might have improved upon 'My heart leaps up when I behold/A rainbow in the sky' (although I have to say it would be hard to improve on the lines that follow).

A further complication is that the raindrops themselves are falling, or blowing about. So any particular raindrop might pass through the band that is delivering, say, red light to you then move into the yellow region. But you continue to see the red band, as if nothing had moved, because new raindrops come to take the places of the departed ones. Richard Whelan, in his lovely *Book of Rainbows* (1997), which is the source of many of my rainbow quotations, quotes Leonardo da Vinci on the subject:

Observe the rays of the sun in the composition of the rainbow, the colours of which are generated by the falling rain, when each drop in its descent takes every colour of the bow.

Treatise on Painting (1490s)

The illusion of the rainbow itself remains rock steady, although the drops that deliver it are falling and scurrying about in the wind. Coleridge wrote,

The steadfast rainbow in the fast-moving, fast-hurrying hail-mist. What a congregation of images and feelings, of fantastic permanence amidst the rapid change of tempest – quietness the daughter of storm.

from *Anima Poetae* (published 1895)

His friend Wordsworth, too, was fascinated by the immobility of the rainbow in the face of turbulent movement of the rain itself:

> *Meanwhile, by what strange chance I cannot tell,*
> *What combination of the wind and clouds,*
> *A large unmutilated rainbow stood*
> *Immovable in heaven.* *The Prelude* (1815)

Part of the romance of the rainbow comes from the illusion that it is always perched on the horizon far away, a huge curve unattainably receding as we approach. But Keats's 'rainbow of the salt sand-wave' was near. And you can sometimes see a rainbow as a complete circle only a few feet in diameter, racing along the near side of a hedge as you drive by. (Rainbows look semicircular only because the horizon gets in the way of the lower part of the circle.) A rainbow looks so big partly because of an illusion of distance. The brain projects the image outwards on to the sky, aggrandizing it. You can achieve the same effect by staring at a bright lamp to 'stamp' its after-image on to your retina, then 'projecting' it into the distance by staring at the sky. This makes it look large.

There are other delightful complications. I said that light from the sun enters a raindrop through the upper quadrant of the surface facing the sun, and leaves through the lower quadrant. But of course there is nothing to stop sunlight entering the lower quadrant. Under the right conditions, it can then be reflected *twice* round the inside of the sphere, leaving the lower quadrant of the drop in such a way as to enter the observer's eye, also refracted, to produce a second rainbow, 8 degrees higher than the first, with the colours reversed. Of course, for any given observer, the two rainbows are delivered by different populations of raindrops. One doesn't often see a double rainbow, but Wordsworth must have done so on occasion, and his heart surely leaped up even higher when he did. Theoretically, there may also be other, yet fainter rainbows arranged concentrically, but they are very seldom seen. Could anyone seriously suggest that it *spoils* it to be told what is going on inside all those thousands of falling, sparkling, reflecting and refracting populations of raindrops? Ruskin said in *Modern Painters III* (1856):

For most men, an ignorant enjoyment is better than an informed one; it is better to conceive the sky as a blue dome than a dark cavity, and the cloud as a golden throne than a sleety mist. I much question whether anyone who knows optics, however religious he may be, can feel in equal degree the pleasure or reverence which an unlettered peasant may feel at the sight of a rainbow . . . We cannot fathom the mystery of a single flower, nor is it intended that we should; but that the pursuit of science should constantly be stayed by the love of beauty, and accuracy of knowledge by tenderness of emotion.

Somehow this all lends plausibility to the theory that poor Ruskin's wedding night was ruined by the horrifying discovery that women have pubic hair.

In 1802, fifteen years before Haydon's 'immortal dinner', the English physicist William Wollaston did a similar experiment to Newton's, but his sunbeam had to pass through a narrow slit before it hit his prism. The spectrum that emerged from the prism was built up as a series of narrow strips of different wavelength. The strips smeared into each other to make a spectrum but, scattered along the spectrum, he saw narrow, dark lines in particular places. These lines were later measured and systematically catalogued by the German physicist Joseph von Fraunhofer, after whom they are now called. The Fraunhofer lines have a characteristic disposition, a fingerprint – or barcode is an even apter analogy – which depends upon the chemical nature of the substance through which the rays have passed. Hydrogen, for example, produces its own characteristic barcode pattern of lines and spaces, sodium a different pattern, and so on. Wollaston saw only seven lines, Fraunhofer's superior instruments revealed 576, and modern spectroscopes about 10,000.

The barcode fingerprint of an element resides not just in the spacing of the lines but in their positioning against the rainbow background. The precise barcodes of hydrogen and all elements are now accurately explained by the quantum theory, but this is where I have to make my excuses and leave. Sometimes I imagine that I have some appreciation of the poetry of quantum theory, but I have yet to achieve an understanding deep enough to explain it to others. Actually, it may be that nobody really understands quantum theory, possibly because natural selection shaped our brains to survive in a world of large, slow things, where quantum effects are smothered. This point is well made by Richard Feynman, who is also supposed to have said, 'If you think you understand quantum theory – you don't understand quantum theory!' I think I have been brought closest to understanding by Feynman's published lectures, and by David Deutsch's astonishing and disturbing book, *The Fabric of Reality* (1997). (I find it additionally disturbing because I cannot tell when I am reading generally accepted physics, versus the author's own daring speculations). Whatever a physicist's doubts about how to interpret quantum theory, nobody doubts its phenomenal success in predicting detailed experimental results. And happily, for the purpose of this chapter, it is enough to know, as we have known since Fraunhofer's time, that each of the chemical elements reliably exhibits a unique barcode of characteristically spaced fine lines, branded across the spectrum.

There are two ways in which Fraunhofer lines may be seen. So far I've mentioned dark lines on a rainbow background. These are caused because an element in the path of light absorbs particular wavelengths, selectively removing them from the rainbow as seen. But an equivalent pattern of bright coloured lines against a dark background is produced if the same element is caused to glow, as when it forms part of the make-up of a star.

Fraunhofer's refinement of Newton's unweaving was already known before the French philosopher Auguste Comte rashly wrote, of the stars:

We shall never be able to study, by any method, their chemical composition or their mineralogical structure . . . Our positive knowledge of stars is necessarily limited to their geometric and mechanical phenomena. *Cours de philosophie positive* (1835)

Today, by meticulous analysis of Fraunhofer barcodes in starlight, we know in great detail what the stars are made of, although our prospects of visiting them are hardly any better than they were in Comte's time. A few years ago, my friend Charles Simonyi had a discussion with a former chairman of the US Federal Reserve Bank. This gentleman was aware that scientists had been surprised when NASA discovered what the moon was really made of. Since the moon was so much closer than the stars, he reasoned, our guesses about the stars are likely to be even more wrong. Sounds plausible but, as Dr Simonyi was able to tell him, the truth is exactly the opposite. No matter how far away the stars may be, they emit their own light, and that makes all the difference. Moonlight is all reflected sunlight (a fact which D. H. Lawrence is said to have refused to believe: it offended his poetic sensibilities), so its spectrum doesn't help us to analyse the moon's chemical nature.

Modern instruments spectacularly outperform Newton's prism, but today's science of spectroscopy is the direct descendant of his unweaving of the rainbow. The spectrum of a star's emitted light, especially its Fraunhofer lines, tells us in great detail which chemical substances are present in a star. It also tells us the temperature, the pressure and the size of the star. It is the basis of an exhaustive classification of the natural history of stars. It puts our sun in its proper place in the great catalogue of stars: a Class G2V Yellow

Dwarf. To quote a popular magazine of astronomy, *Sky and Telescope*, from 1996:

To those who can read its meaning, the spectral code tells at a glance just what kind of object the star is – its color, size, and luminosity, its history and future, its peculiarities, and how it compares with the Sun and stars of all other types.

By unweaving starlight in spectroscopes we know that stars are nuclear furnaces, fusing helium out of the hydrogen that dominates their mass; then thrusting helium nuclei together in the further cascade of impurities which make up most of the rest of the elements, forging the medium-sized atoms of which we are eventually made.

Newton's unweaving paved the way for the nineteenth-century discovery that the visible rainbow, the band that we actually see, is a narrow chink in the full spectrum of electromagnetic waves. Visible light spans the wavelengths from 0.4 millionths of a metre (violet) to 0.7 millionths of a metre (deep red). A little longer than red are infrared rays, which we perceive as invisible heat radiation and which some snakes and guided missiles use to home in on their targets. A little shorter than violet are ultraviolet rays, which burn our skin and give us cancer. Radio waves are much longer than red light. Their wavelengths are measured in centimetres, metres, even thousands of metres. Between them and infrared waves on the spectrum lie microwaves, which we use for radar and for high-speed cooking. Shorter than ultraviolet rays are X-rays, which we use for seeing bone through flesh. Shortest of all are cosmic rays, with a wavelength measured in trillionths of a metre. There is nothing special about the narrow band of wavelengths that we call light, apart from the fact that we can see it. For insects, visible light is shifted bodily along the spectrum. Ultraviolet is for them a visible colour ('bee purple'), and they are blind to red (which they might call 'infra yellow'). Radiation all along the larger spectrum can be unwoven in the same kind of way as the rainbow, although the particular instrument we use for the unweaving – a radio tuner instead of a prism, for instance – is different in different parts of the spectrum.

The colours that we actually experience, the subjective sensations of redness and blueness, are arbitrary labels that our brains tie to light of different wavelengths. There is nothing intrinsically 'long' about redness. Knowing how red and blue look doesn't help us remember which wavelength is longer. I regularly have to look it up, whereas I never forget that soprano sounds have a shorter wavelength than bass. The brain needs convenient internal labels for the different parts of the physical rainbow. Nobody knows if my sensation of redness matches yours, but we can easily agree that the light that I call red is the same as the light that you call red and that, if a physicist measures it, it will be found to have a long wavelength. My subjective judgement is that violet looks redder than blue does, even though it lies further away on the spectrum from red. Probably you agree. The apparent reddish tinge in violet is a fact about nervous systems, not a fact about the physics of spectrums.

Hugh Lofting's immortal Doctor Dolittle flew to the moon and was startled to see a dazzling range of new colours, as different from our familiar colours as red is from blue. Even in fiction we can be sure that this would never happen. The hues that would greet any traveller to another world would be a function of the brains that they bring with them from the home planet.*

* Colour is a rich source of philosophical speculation, which is often scientifically under-informed. A laudable attempt to rectify this is C. L. Hardin's 1988 book, *Color for Philosophers: Unweaving the Rainbow*. I am embarrassed to say that I discovered this book, and in particular its excellent subtitle, only after mine had gone off to the publishers. Doctor Dolittle, by the way, may be hard to find, as he is now often banned by pompously correct librarians. They worry about the racism in *The Story of Doctor Dolittle*, but this was all but universal in the 1920s. In any case it is offset by the Doctor's splendid fight against the slave trade in *Doctor Dolittle's Post Office*, and, more profoundly, by the stand that all the Doctor Dolittle books make against the vice of speciesism which is as unquestioned today as racism was in earlier times.

We know now in some detail how the eye informs the brain about the wavelengths of light. It is a three-colour code, as used in colour television. The human retina has four kinds of light-sensitive cell: three kinds of 'cones' plus the 'rods'. All four are similar and have surely diverged from a common ancestor. One of the things it is easy to forget about any sort of cell is how intricately complicated even a single cell is, much of the complexity being built up of fine-folded internal membranes. Each tiny rod or cone contains a deep stack of membranes, packed like a tall column of books. Threaded back and forth through each book is a long, thin molecule, a protein called retinal. Like many proteins, retinal behaves as an enzyme, catalysing a particular chemical reaction by providing a correctly shaped place for particular molecules to slot in.

It is the three-dimensional form of an enzyme molecule which gives it its catalytic property, serving as a carefully shaped, albeit slightly flexible, template for other molecules to fit in and meet one another – otherwise they'd have to rely on bumping into each other by chance (which is why enzymes so dramatically speed up chemical reactions). The elegance of this system is one of the key things that makes life possible, but it does raise a problem. Enzyme molecules are often capable of coiling into more than one shape, and usually only one of them is desirable. Much of the work of natural selection over the millions of years has been to find 'decisive' or 'single-minded' molecules whose 'preference' for their favoured shape is much stronger than their tendency to coil into any other shape. Molecules with two alternative shapes can be a tragic menace. 'Mad cow disease', sheep scrapie and their human counterparts Kuru and Creutzfeldt-Jakob disease, are caused by proteins called prions which have two alternative shapes. They normally fold into one of the two shapes, and in this configuration they do a useful job. But occasionally they adopt the alternative shape. And then a terrible thing happens. The presence of one protein with the alternative shape induces others to come over to the rogue persuasion. An epidemic of misshapen proteins sweeps through the body like a cascade of falling dominoes. A single misshapen protein can infect a new body and trigger a new domino run. The consequence is death from spongy holes in the brain, because the protein in its alternative shape cannot do its normal job.

Prions have caused some confusion because they spread like self-replicating viruses, yet they are proteins, and proteins aren't supposed to be self-replicating. Textbooks of biology would have it that self-replication is the unique privilege of polynucleotides (DNA and RNA). However, prions are self-replicating only in the peculiar sense of one misshapen rogue molecule 'persuading' its already existing neighbours to flip into the same shape.

In other cases, enzymes with two alternative shapes turn their switchability to good account. Switchability is, after all, the essential property of transistors, diodes and the other high-speed electronic gates that make the logical operations of computers – IF, NOT, AND, OR and the like – possible. There are 'allosteric' proteins that flip from state to state in a transistor-like way, not through infectious 'persuasion' by a neighbour, as in prions, but only IF some biologically useful condition is met, AND NOT under certain other conditions. Retinal is one of these 'transistor' proteins which make good use of their property of having two alternative shapes. Like a photocell, it flips from one state to the other when it is hit by light. It automatically clicks back to the previous shape after a brief recovery period. In one of its two shapes it is a powerful catalyst, but not in the other. So, when light causes it to snap into its active shape, this initiates a special chain reaction and a rapid turnover of molecules. It is as if the light had switched on a high-pressure tap.

The end product of the resulting chemical cascade is a stream of nerve impulses which are relayed to the brain via a series of nerve cells, each of which is a long thin tube. Nerve impulses, too, are rapidly catalysed chemical changes. They sweep along the long thin tubes like fizzing trails of gunpowder. Each fizzing sweep is discrete and separate from the others, so they arrive at the far end of the tube like a series of short, sharp reports – nerve impulses. The rate at which the nerve impulses arrive – which may be hundreds per second – is a coded representation of (in this case) the intensity of light falling on the rod or cone cell. As far as a single nerve cell is concerned, the difference between strong stimulation and weak is the difference between a high-speed machine gun and intermittent rifle fire.

So far, what I have said applies to the rods and all three kinds of cone. Now for the ways in which they differ. Cones respond only to bright light. Rods are sensitive to dim light and are needed for night vision. Rods are found all over the retina, and are nowhere particularly crowded, so they are no good for resolving small detail. You can't read with them. You read with cones, which are extremely densely packed in one particular part of the retina, the fovea. The denser the packing, of course, the finer the detail that can be resolved.

Rods are not involved in colour vision because they all have the same wavelength sensitivity as each other. They are most sensitive to yellow light in the middle of the visible spectrum, less sensitive nearer the two ends of the spectrum. This does not mean that they report all light to the brain as yellow. That isn't even a meaningful thing to say. All nerve cells report to the brain as nerve impulses, and that's all. If a rod fires rapidly, this could either mean that there is a lot of red or blue light, or that there is somewhat less yellow light. The only way for the brain to resolve the ambiguity is to have simultaneous reports from more than one kind of cell, differentially sensitive to different colours.

This is where the three kinds of cone come in. The three kinds of cone have three different flavours of retinal. All of them respond to light of all wavelengths. But one kind is most sensitive to blue light, another is most sensitive to green light, and the third is most sensitive to red light. By comparing the firing rates of the three kinds of cone – in effect, subtracting them from each other – the nervous system is able to reconstruct the wavelengths of light hitting the relevant part of the retina. Unlike the case of vision by rods alone, the brain is not confused between dim light of one colour and bright light of another colour. The brain, because it receives reports from more than one kind of cone, is able to compute the true colour of the light.

As I said when recalling Doctor Dolittle on the moon, the colours that we finally think we see are labels used for convenience by the brain. I used to be disappointed when I saw 'false colour' images, say, satellite photographs of earth, or computer-constructed images of deep space. The caption tells us that the colours are arbitrary codes, say, for different types of vegetation, in a satellite picture of Africa. I used to think false colour images were a kind of cheat. I wanted to know what the scene 'really' looked like. I now realize that everything I think I see, even the colours of my own garden through the window, are 'false' in the same sense: arbitrary conventions used, in this case by my brain, as convenient labels for wavelengths of light. Chapter 11 argues that all our perceptions are a kind of 'constrained virtual reality' constructed in the brain. (Actually, I am *still* disappointed by false colour images!)

We can never know whether the subjective sensations that different people associate with particular wavelengths are the same. We can compare opinions about what colours seem to be mixtures of which. Most of us agree to find it plausible that orange is a mixture of red and yellow. Blue-green's status as a mixture is conveyed by the compound word itself, though not by the word turquoise. It is controversial whether different languages agree on how they partition the spectrum. Some linguists aver that the Welsh language does not divide the green and blue region of the spectrum in the same way as English does. Instead, Welsh is said to have a word corresponding to part of green, and another word corresponding to the other part of green plus part of blue. Other linguists and anthropologists say that this is a myth, no more true than the equally seductive allegation that the Inuit ('Eskimos') have 50 different words for snow. These sceptics claim experimental evidence, obtained by presenting a large range of coloured chips to native speakers of many languages, that there are strong universals in the way humans partition the spectrum. Experimental evidence is, indeed, the only way to settle the question. It matters nothing that, at least to this English speaker, the story about the Welsh partitioning of blue and green feels implausible. There is nothing in physics to gainsay it. The facts, whatever they are, are facts of psychology.

Unlike birds, which have excellent colour vision, many mammals have no true colour vision at all. Others, including certain kinds of partially colour-blind humans, use a two-colour system based on two kinds of cones. High-quality colour vision with a three-colour system may have evolved in our primate ancestors as an aid to finding fruits in the green forest. It has even been suggested, by the Cambridge psychologist John Mollon, that the three-colour system 'is a device invented by certain fruiting trees in order to propagate themselves': an imaginative way of calling attention to the fact that trees benefit from attracting mammals to eat their fruits and spread the seeds. Some New World monkeys go in for weird arrangements in which different individuals within a species have different combinations of two-colour systems, and are thereby specialized to see different things. Nobody knows whether or how this benefits them, but it may be suggestive that bomber crews in the Second World War liked to include at least one colour-blind member, who could penetrate certain kinds of camouflage on the ground.

Unweaving the wider rainbow, moving to other parts of the electromagnetic spectrum, we separate station from station on the radio dial, we insulate conversation from conversation in the network of cellular telephones. Without sensitive unweaving of the electromagnetic rainbow, we'd hear everybody's conversation simultaneously, and all the frequencies on the broadcasting dial, in a white babel of noise. In a different way, and with the assistance of special-purpose computers, unweaving the rainbow underlies

Magnetic Resonance Imaging, the spectacular technique by which doctors today can discern the three-dimensional structure of our internal organs.

When a source of waves is itself moving relative to its detector, something special happens. There is a 'Doppler shift' of wavelengths as detected. This is easy to notice in the case of sound waves because they travel slowly. A car's engine note is of a distinctly higher pitch when it is approaching than when receding. This is why we hear the characteristic dual-tone eee-aaah when a car whizzes past. The Dutch scientist Buys Ballott in 1845 first verified Doppler's prediction by hiring a brass band to play on an open railway wagon as it sped past his audience. Light waves travel so fast that we notice the Doppler shift only if we are moving very fast towards the source of light (in which case the light is shifted towards the blue end of the spectrum) or away from it (in which case the light is red-shifted). This is true of distant galaxies. The fact that they are fast receding from us was first discovered because of the Doppler shift in their light. It is redder than it should be, shifted consistently towards the longwave, red end of the spectrum.

How do we know that the light coming from a distant galaxy is red-shifted? How do we know that it wasn't red when it set out? You can tell by using the Fraunhofer lines as markers. Each element, remember, signs its name in a unique barcode of lines. The spacing between the lines is characteristic like a fingerprint, but so is the precise position of each line along the rainbow. Light from a distant galaxy shows barcodes that have familiar spacing patterns. This very familiarity is what tells us that other galaxies are made of the same range of stuffs as our own. But the whole pattern is shifted a fixed distance towards the longwave end of the spectrum: it is redder than it should be. In the 1920s, the American astronomer Edwin Hubble (after whom the Hubble Space Telescope is now named) discovered that distant galaxies have red-shifted spectra. Those galaxies with the most pronounced red shift are also the most distant – as estimated from the faintness of their light.

Hubble's famous conclusion (although it had been suggested by others before) was that the universe is expanding, and from any given observation point the galaxies seem to recede at ever-increasing speed.

When we look at a distant galaxy, we are looking far back into the past, for the light has taken billions of years to reach us. It has become faint, which is how we know it has come a great distance. The speed with which our galaxy is racing apart from the other galaxy has had the effect of shifting the spectrum towards the red end. The relationship between distance and velocity of receding is a lawful one (it obeys 'Hubble's law'). By extrapolating this quantitative relationship backwards we can estimate when the universe began expanding. Using the language of the now prevailing 'Big Bang' theory, the universe began in a gigantic explosion between ten billion and 20 billion years ago. All this we infer from unweaving the rainbow. Further developments of the theory, supported by all available evidence, suggest that time itself began in this mother of all cataclysms. You probably don't understand, and I certainly don't, what it can possibly mean to say that time itself began at a particular moment. But once again that is a limitation of our minds, which were only ever designed to cope with slow, rather large objects on the African savannahs, where events come well behaved and in order, and every one has a before. An event that has no before terrifies our poor reason. Maybe we can appreciate it only through poetry. Keats, thou shouldst be living at this hour.

And are there eyes out there in the galaxies, looking back at us? Back is the word, for they can see us only in our past. The inhabitants of a world 100 million light years distant might at this moment see, if they could see anything at all on our planet, red-shifted dinosaurs lunging over the rose tinted plains. Alas, even if there are other creatures in the universe, and even if they have eyes, it is unlikely that, however powerful their telescopes, they will have the resolving power to see our planet, let alone individual inhabitants of it. We ourselves have never seen another planet outside our solar system. We didn't even know about all the planets in our own solar system until recent centuries. Neptune and Pluto are too faint to be seen by the naked eye. The only reason we knew which way to point the telescope was by calculations from minute perturbations in the orbits of nearer planets. In 1846, two mathematical astronomers, J. C. Adams in England and U. J. J. Leverrier in France, were independently puzzled by a discrepancy between the actual position of the planet Uranus and where it theoretically should have been. Both calculated that the perturbation could have been caused by the gravity of an invisible planet of a particular mass in a particular place. The German astronomer J. G. Galle duly pointed his telescope in the right direction and discovered Neptune. Pluto was discovered in the same way, as late as 1930 by the American astronomer C. W. Tombaugh, alerted by its (much smaller) gravitational effects on the orbit of Neptune. John Keats would have appreciated the excitement those astronomers felt:

> *Then felt I like some watcher of the skies*
> *When a new planet swims into his ken;*
> *Or like stout Cortez when with eagle eyes*
> *He stared at the Pacific – and all his men*
> *Look'd at each other with a wild surmise –*
> *Silent upon a peak in Darien.*
> 'On First Looking into Chapman's Homer' (1816)

I have had a special affection for these lines ever since they were quoted to me by a publisher on first reading the manuscript of *The Blind Watchmaker.*

But are there planets orbiting other stars? An important question, this, whose answer affects our estimate of the ubiquity of life in the universe. If there is only one star in the universe that has planets, that star has to be our sun and we are very very alone. At the other extreme, if every star is the centre of a solar system, the number of planets potentially available for life will exceed all counting. Almost whatever the odds of life on any one planet, if we find planets orbiting one other typical star as well as the sun, we feel sensibly less lonely.

Planets are too close to their suns, and too smothered by their brightness, for our telescopes normally to see them. The main way we know that other stars have planets – and the discovery waited till the 1990s – is, once again, through orbital perturbations, this time detected via Doppler shifts in coloured light. Here's how this works. We think of the sun as the centre about which planets orbit. But Newton tells us that two bodies orbit each other. If two stars are of similar mass – they're called a binary pair – the two swing round each other like a pair of dumb-bells. The more unequal they are, the more it seems the lighter one orbits the heavier one, which almost stays still. When one body is much larger than the other, for instance the sun versus Jupiter, the heavier one just wobbles a bit while the lighter one whizzes round like a terrier circling its master on a walk.

It is such wobbles in the positions of stars that betray the presence of otherwise invisible planets orbiting them. But the wobbles themselves are too small to be seen directly. Our telescopes cannot resolve such small changes in position; less so, indeed, than they can resolve the planets themselves. Again, it is unweaving the rainbow that comes to the rescue. As a star wobbles back and forth under the influence of an orbiting planet, the light from it reaches us red-shifted when the star is moving away, blue-shifted when it is moving towards us. Planets give themselves away by causing minute, but measurable, red/blue oscillations in the light reaching us from their parent stars. In the same way, inhabitants of distant planets might just detect the presence of Jupiter by watching the sun's rhythmic changes of hue. Jupiter is probably the only one of our sun's planets large enough to be detectable in this way. Our humble planet is too tiny to make gravitational ripples for aliens to notice.

They might, however, be aware of us through unweaving the rainbow of radio and television signals that we ourselves have been pumping out for the past few decades. The swelling spherical bubble of vibrations, now more than a light-century across, has enveloped a significant number of stars though an insignificant proportion of those that populate the universe. Carl Sagan, in his novel *Contact*, has darkly noted that in the vanguard of images announcing earth to the rest of the universe will be Hitler's speech opening the 1936 Olympic Games in Berlin. No reply has so far been picked up, no message of any kind from any other world.

We have never been given any direct reason to suppose that we have company. In very different ways, the possibility that the universe is teeming with life, and the opposite possibility that we are totally alone, are equally exciting. Either way, the urge to know more about the universe seems to me irresistible, and I cannot imagine that anybody of truly poetic sensibility could disagree. I am ironically amused by how much of what we have discovered so far is a direct extrapolation of unweaving the rainbow. And the poetic beauty of what that unweaving has now revealed, from the nature of the stars to the expansion of the universe, could not fail to catch the imagination of Keats; would be bound to send Coleridge into a frenzied reverie; would make Wordsworth's heart leap up as never before.

The great Indian astrophysicist Subrahmanyan Chandrasekhar said, in a lecture in 1975:

This 'shuddering before the beautiful', this incredible fact that a discovery motivated by a search after the beautiful in mathematics should find its exact replica in Nature, persuades me to say that beauty is that to which the human mind responds at its deepest and most profound.

How much more sincere that sounds than Keats's better-known expression of a superficially similar emotion:

> *'Beauty is truth, truth beauty,' – that is all*
> *Ye know on earth, and all ye need to know.*
>
> 'Ode on a Grecian Urn' (1820)

Keats and Lamb should have raised their glass to poetry, and to mathematics, and to the poetry of mathematics. Wordsworth would have needed no encouragement. He (and Coleridge) had been inspired by the Scottish poet James Thomson, and might have recalled Thomson's 'To the Memory of Sir Isaac Newton' (1727):

> . . . *Even Light itself, which every thing displays,*
> *Shone undiscovered, till his brighter mind*
> *Untwisted all the shining robe of day;*
> *And, from the whitening undistinguished blaze,*
> *Collecting every ray into his kind,*
> *To the charmed eye educed the gorgeous train*
> *Of parent colours. First the flaming red*
> *Sprung vivid forth; the tawny orange next;*
> *And next delicious yellow; by whose side*
> *Fell the kind beams of all-refreshing green.*
> *Then the pure blue, that swells autumnal skies,*
> *Ethereal played; and then, of sadder hue,*
> *Emerged the deepened indigo, as when*
> *The heavy-skirted evening droops with frost;*
> *While the last gleamings of refracted light*
> *Died in the fainting violet away.*
> *These, when the clouds distil the rosy shower,*
> *Shine out distinct adown the watery bow;*
> *While o'er our heads the dewy vision bends*
> *Delightful, melting on the fields beneath.*
> *Myriads of mingling dyes from these result,*
> *And myriads still remain – infinite source*
> *Of beauty, ever flushing, ever new.*

Did ever poet image aught so fair,
Dreaming in whispering groves by the hoarse brook?
Or prophet, to whose rapture heaven descends?
Even now the setting sun and shifting clouds,
Seen, Greenwich, from thy lovely heights, declare
How just, how beauteous the refractive law.

4
BARCODES ON THE AIR

We shall find the Cube of the Rainbow,
Of that, there is no doubt.
But the Arc of a Lover's conjecture
Eludes the finding out.

EMILY DICKINSON (1894)

On the air, in contemporary English, means on the radio. But radio waves have nothing to do with air, they are better regarded as invisible light waves with long wavelengths. Airwaves can sensibly mean only one thing and that is sound. This chapter is about sound and other slow waves, and how they, too, can be unwoven like a rainbow. Sound waves travel half a million times more slowly than light (or radio) waves, not much faster than a Boeing 747 and slower than a Concorde. Unlike light and other electromagnetic radiation, which propagates best through a vacuum, sound waves travel only through a material medium such as air or water. They are waves of compression and rarefaction (thickening and thinning) of the medium. In air, this means waves of increasing and then decreasing local barometric pressure. Our ears are tiny barometers capable of tracking high-speed rhythmic changes of pressure. Insect ears work in another way entirely. In order to understand the difference, we need a small digression to examine what pressure really is.

We feel pressure on our skin, when we place our hand over the outlet of a bicycle pump, for example, as a kind of springy push. Actually, pressure is the summed bombardments of thousands of molecules of air, whizzing about in random directions (as opposed to a wind, where the molecules predominantly flow in one particular direction). If you hold your palm up to a high wind you feel the equivalent of pressure – bombardment of molecules. The molecules in a confined space, say, the interior of a well-pumped bicycle tyre, press outwards on the walls of the tyre with a force proportional to the number of molecules in the tyre and to the temperature. At any temperature higher than −273°C (the lowest possible temperature, corresponding to complete motionlessness of molecules), the molecules are in continuous random motion, bouncing off each other like billiard balls. They don't only bounce off each other, they bounce off the inside walls of the tyre – and the walls of the tyre 'feel' it as pressure. As an additional effect, the hotter the air, the faster the molecules rush about (that's what temperature means), so the pressure of a given volume of air goes up when you heat it. By the same token, the temperature of a given quantity of air goes up when you compress it, i.e. raise the pressure by reducing the volume.

Sound waves are waves of oscillating local pressure change. The total pressure in, say, a sealed room is determined by the number of molecules in the room and the temperature, and these numbers don't change in the short term. On average, every cubic centimetre in the room will have the same number of molecules as every other cubic centimetre, and therefore the same pressure. But this doesn't stop there being local variations in pressure. Cubic centimetre A may experience a momentary rise in pressure at the expense of cubic centimetre B, which has temporarily donated some molecules to it. The increased pressure in A will tend to push molecules back to B and redress the balance. On the much larger scale of geography, this is what winds are – flows of air from high-pressure areas to low-pressure areas. On a smaller scale sounds can be understood in this way, but they are not winds because they oscillate backwards and forwards very fast.

If a tuning fork is struck in the middle of a room, the vibration disturbs the local molecules of air, causing them to bump into neighbouring molecules of air. The tuning fork vibrates back and forth at a particular frequency, causing ripples of disturbance to propagate outwards in all directions as a series of expanding shells. Each wavefront is a zone of increased pressure, with a zone of decreased pressure following in its wake. Then the next wavefront comes, after an interval determined by the rate at which the tuning fork is vibrating. If you stick a tiny, very fast-acting barometer anywhere in the room, the barometer needle will swing up and down as each wavefront passes over it. The rate at which the barometer needle oscillates is the frequency of the sound. A fast-acting barometer is exactly what a vertebrate ear is. The eardrum moves in and out under the changing pressures that hit it. The eardrum is connected (via three tiny bones, the famous hammer, anvil and stirrup, sequestered in evolution from the bones of the reptilian jaw hinge) to a kind of inverse harp in miniature, called the cochlea. As in a harp, the 'strings' of the cochlea are arranged across a tapering frame. Strings at the small end of the frame vibrate in sympathy with high-pitched sounds, those at the big end vibrate in sympathy with low-pitched sounds. Nerves from along the cochlea are mapped in an orderly way in the brain, so the brain can tell whether a low-pitched or a high-pitched sound is vibrating the eardrum.

Insect ears, by contrast, are not little barometers, they are little weathervanes. They actually measure the flow of molecules as a wind, albeit a queer kind of wind which travels only a very short distance before reversing its direction. The expanding wavefront which we detect as a change in pressure is also a wave of movement of molecules: movement into a local area as the pressure goes up, then movement back out of that area as the pressure goes down again. Whereas our barometer ears have a membrane stretched over a confined space, insect weathervane ears have either a hair, or a membrane stretched over a chamber with a hole. In either case, it is literally blown back and forth by the rhythmic, backward and forward movements of the molecules.

Sensing the direction of a sound is therefore second nature for insects. Any fool with a weathervane can distinguish a north wind from an east wind, and a single insect ear finds it easy to tell a north–south oscillation from an east–west oscillation. Directionality is built into insects' method of detecting sound. Barometers aren't like that. A rise in pressure is just a rise in pressure, and it doesn't matter from which direction the added molecules come. We vertebrates therefore, with our barometer ears, have to calculate the direction of sound by comparing the reports of the two ears, rather as we calculate colour by comparing the reports of different classes of cones. The brain compares the loudness at the two ears and separately it compares the time of arrival of sounds (especially staccato sounds) at the two ears. Some kinds of sounds lend themselves to such comparisons less readily than others. Cricket song is cunningly pitched and timed so as to be hard for vertebrate ears to locate, but easy for female crickets, with their weathervane ears, to home in upon. Some cricket chirps even create the illusion, at least to my vertebrate brain, that the (in fact stationary) cricket is leaping about like a jumping squib.

Sound waves form a spectrum of wavelengths, analogous to the rainbow. The sound rainbow is also subject to unweaving, which is why it is possible to make any sense of sounds at all. Just as our sensations of colour are the labels that the brain slaps on to light of different wavelengths, the equivalent internal labels that it uses for sounds are the different pitches. But there is a lot more to sound than simple pitch, and this is where unweaving really comes into its own.

A tuning fork or a glass harmonica (an instrument favoured by Mozart, made from fine glass bowls tuned by the depth of water they contain, sounded by a wetted finger drawn around the rim) emit a crystalline pure sound. Physicists call these sine waves. Sine waves are the simplest kind of waves, sort of theoretical ideal waves. The smooth curves that snake along a rope when you wiggle one end up and down are more or less sine waves, although of much lower frequency than sound waves, of course. Most sounds are not simple sine waves but are more jagged and complicated, as we shall see. For the moment we shall think of a tuning fork or glass harmonica, singing out its smooth, curvaceous waves of pressure change that race away from the source in concentrically expanding spheres. A barometer ear placed at one spot detects a smooth increase in pressure followed by a smooth decrease, rhythmically oscillating with no kinks or wiggles in the curve. With every doubling in frequency (or halving in wavelength, which is the same thing) we hear a jump of one octave. Very low frequencies, the deepest notes of the organ, shudder through our bodies and are hardly heard by our ears at all. Very high frequencies are inaudible to humans (especially older humans) but audible to bats and used by them, in the form of echoes, to find their way about. This is one of the most enthralling stories in all natural history, but I've devoted a whole chapter to it in *The Blind Watchmaker* so will resist the temptation to expand.

Tuning forks and glass harmonicas aside, pure sine waves are largely a mathematical abstraction. Real sounds are mostly more complicated mixtures, and they richly repay unweaving. Our brains unweave them effortlessly and to astonishing effect. It is only with much labour that our mathematical understanding of what is going on has caught up, clumsily and incompletely, with what our ears have effortlessly unwoven – and our brains rewoven – from childhood on.

Suppose we sound one tuning fork with an oscillating frequency of 440 cycles per second, or 440 Herz (Hz). We shall hear a pure tone, the A above middle C. What is the difference between this and a violin playing the same A, a clarinet playing the same A, an oboe, a flute? The answer is that each instrument includes admixtures of waves whose frequencies are various multiples of the fundamental frequency. Any instrument playing the A above middle C will deliver most of its sound energy at the fundamental frequency, 440 Hz, but superimposed will be traces of vibration at 880 Hz, 1320 Hz and so on. These are called harmonics, although the word can be confusing since 'harmonies' are chords of several notes that we hear as distinct. A 'single' trumpet note is actually a mixture of harmonics, the particular mixture being a kind of trumpet 'signature' that distinguishes it from, say, a violin playing the 'same' note (with different, violin signature harmonics). There are additional complications, which I shall ignore, around the onset of sounds, for example the lippy irruption of a trumpet blast or the zing as a violin bow hits the string.

These complications aside, there is a characteristic trumpet (or violin, or whatever it is) quality to the sustained part of a note. It is possible to demonstrate that the apparently single tone of a particular instrument is a rewoven construct of the brain, summing up sine waves. The demonstration works as follows. Having decided which sine waves are involved in, say, trumpet sound, select the appropriate 'tuning fork' pure tones and sound them one at a time. For a brief period you can hear the separate notes, as if they really were a chord of tuning forks. Then, quite eerily, they click into focus with each other, the 'tuning forks' disappear, and you hear only what Keats called the silver, snarling trumpets, sounding the pitch of the fundamental frequency. A different barcode combination of frequencies is needed to make the sound of a clarinet, and again you can fleetingly distinguish them as separate 'tuning forks' before the brain homes in on the illusion of one 'woody' clarinet note. The violin has its own barcode signature, and so on.

Now, if you watch a tracing of the pressure wave when a violin is playing some note, what you see is a complicated wiggly line repeating itself at the fundamental frequency but with smaller wiggles of higher frequency superimposed. What has happened is that the different sine waves that constitute violin noise have summed up to make the complicated wiggly line. It is possible to program a computer to analyse any complicatedly repeating pattern of wiggles back into its component pure waves, the separate sine waves that you would have to sum up to make the complicated pattern. Presumably, when you listen to an instrument, you are performing something equivalent to this calculation, the ear first unweaving the component sine waves, then the brain weaving them together again and giving them the appropriate label: 'trumpet', 'oboe' or whatever it is.

But our unconscious feats of unweaving and weaving are greater even than this. Think what is happening when you listen to a whole orchestra. Imagine that, superimposed on a hundred instruments, your neighbour in the concert is whispering learned music criticism in your ear, others are coughing and, lamentably, somebody behind you is rustling a chocolate wrapper. All these sounds, simultaneously, are vibrating your eardrum and they are summed into a single, very complicated wriggling wave of pressure change. We know it is one wave because a full orchestra, and all the noises off, can be rendered into a single wavy groove on a phonograph disc, or a single fluctuating trace of magnetic substance on a tape. The entire set of vibrations sums up into a single wiggly line on the graph of air pressure against time, as recorded by your eardrum. *Mirabile dictu*, the brain manages to sort out the rustling from the whispering, the coughing from the door banging, the instruments of the orchestra from each other. Such a feat of unweaving and reweaving, or analysis and synthesis, is almost beyond belief, but we all do it effortlessly and without thinking. Bats are even more impressive, analysing stuttering volleys of echoes to build up, in their brains, detailed and fast-changing three-dimensional images of the world through which they fly, including the insects which they catch on the wing, and even sorting out their own echoes from those of other bats.

The mathematical technique of decomposing wiggling waveforms into sine waves which can then be summed again to make the original wiggly line is called Fourier analysis, after the nineteenth-century French mathematician Joseph Fourier. It works not just for sound waves (indeed, Fourier himself developed the technique for a quite different purpose) but for any process that varies periodically, and it doesn't have to be high-speed waves like sound, or ultra-high-speed waves like light. We can think of Fourier analysis as a mathematical technique which is convenient for unweaving 'rainbows' where the vibration that makes up the spectrum is slow compared with that of light.

To go to a very slow vibration indeed, I recently saw, on a road in the Kruger National Park in South Africa, a wiggly wet line which followed the course of the road and apparently traced out some kind of complicated repeat pattern. My host and expert guide told me that it was a trail of urine from a male elephant in musth. When a bull elephant enters this curious state (perhaps the elephantine equivalent of an Australian on 'walkabout') he dribbles out urine more or less continuously, apparently for scent-marking purposes. The side-to-side waving of the urine trail on the road was presumably produced by the long penis acting as a pendulum (it would be a sine wave if the penis were a perfect, Newtonian pendulum, which it is not) interacting with the more complicated periodicity of the lumbering four-footed gait of the whole animal. I took photographs with the vague intention of later performing a Fourier analysis. I am sorry to say I have never got around to doing it. But in theory it could be done. A tracing of the photographed urine line could be laid over squared paper and its coordinates digitized for feeding into a computer. The computer could then perform a modern version of Fourier's calculations and extract the component sine waves. There are easier (though not necessarily safer) ways to measure the length of an elephant's penis, but it would have been fun to do, and Baron Fourier himself would surely have been delighted at such an unexpected use of his mathematics. There is no reason why a urine trail might not fossilize, as footprints and wormcasts do, in which case we could in principle use Fourier analysis to measure the penis length of an extinct mastodon or woolly mammoth, from the indirect evidence of its urine trail in musth.

An elephant's penis swings at a frequency much slower than sound (although in the same ballpark as sound when you compare it with the ultra high frequencies of light). Nature offers us other periodic waveforms, of much lower frequency still, with wavelengths measured in years or even millions of years. Some of these have been subjected to the equivalent of Fourier analysis, including the cycles of animal populations. Since 1736, the Hudson's Bay Company kept records of the abundance of pelts brought in by Canadian fur trappers. The distinguished Oxford ecologist Charles Elton (1900–1991), who was employed as a consultant by the company, realized that these records could provide a read-out of fluctuating populations of snowshoe hares, lynxes and other mammals persecuted by the fur trade. The figures rise and fall in complicated mixtures of rhythms, which have been much analysed. Among the wavelengths that have been pulled out by these analyses is a prominent one of approximately four-year periodicity, and another of around 11 years. One hypothesis that has been suggested to account for the four-year rhythms is a time-lagged interaction between predators and prey (a glut of prey feeds a plague of predators, who then nearly wipe out the prey; this in turn starves the predators, then the consequent drop in predator population allows a new boom in the prey population, and so on). As for the longer rhythm of 11 years, perhaps the most intriguing suggestion connects it with sunspot activity, which is known to vary on an approximately 11-year cycle. How the sunspots affect animal populations is open to discussion. Perhaps they change the earth's weather, which affects abundance of plant food.

Wherever you find regular cycles of very long wavelengths, they are likely to have astronomical origins. They stem from the fact that celestial objects often rotate on their own axis, or follow repetitious orbits around other celestial objects. Twenty-four-hour rhythms of activity pervade almost all the fine details of living bodies on this planet. The ultimate reason is the rotation of the earth about its own axis, but animals of many species, including humans, when isolated from direct contact with day and night, continue to cycle on with a rhythm of approximately 24 hours, showing that they have internalized the rhythm and can free-run it even in the absence of the external pacemaker. The lunar rhythm of 28 days is another prominent component of the mix of waves in the bodily functions of many creatures, especially marine ones. The moon exerts its rhythmic influence via the succession of spring and neap tides. The earth's orbital rhythm of slightly more than 365 days contributes its slower pendulum to the Fourier sum, manifesting itself via breeding seasons, seasons of migration, patterns of moulting and growth of winter coats.

Perhaps the longest wavelength picked up by the unweaving of biological rhythms is a suggested 26-million-year cycle of mass extinctions. Fossil experts reckon that more than 99 per cent of the species that have ever lived have become extinct. Fortunately, the rate of extinction is, over the long term, roughly balanced by the rate at which new species are formed by the splitting of existing ones. But this doesn't mean they stay constant in the shorter term. Far from it. Extinction rates fluctuate all over the place, and so do the rates at which new species come into existence. There are bad times when species disappear, and good times when they burgeon. Probably the worst of the bad times, the most devastating Armageddon, occurred at the end of the Permian era, about a quarter of a billion years ago. Around 90 per cent of all species became extinct in that terrible time, including on land many mammal-like reptiles. Earth's fauna eventually bounced back on to the denuded stage, but with a very different cast list: on land the dinosaurs stepped into the range of costumes left by dead mammal-like reptiles. The next largest mass extinction – and the most talked-about – is the famous Cretaceous extinction of 65 million years ago, in which all the dinosaurs, and many other species with them both on land and in the sea, were wiped out, instantaneously as far as the fossil record can tell. In the Cretaceous event, perhaps 50 per cent of all species went extinct, not as many as in the Permian but nevertheless this was a fearful global tragedy. Once again, our planet's devastated fauna bounced back and here we are, we mammals, descended from a few fortunate relicts of the once rich mammal-like reptile fauna. Now we, together with the birds, fill gaps left by the dead dinosaurs. Until, presumably, the next great extinction.

There have been many episodes of mass extinction, not as bad as the Permian and Cretaceous events, but still noticeable in the chronicles of the rocks. Statistical paleontologists have gathered the numbers of fossil species over the ages and fed them into computers to perform Fourier analysis and extract such rhythms as they can find, as if listening for the flutter of preposterously deep organ notes. The dominant rhythm that has been claimed (albeit controversially) is a periodicity of about 26 million years. What could cause rhythms of extinction with such a formidably long wavelength? Probably only a celestial cycle.

Evidence is accumulating that the Cretaceous catastrophe was caused when a large asteroid or comet, the size of a mountain and travelling at tens of thousands of miles per hour, scored a direct hit on our planet, probably somewhere around what we now call the Yucatán peninsula in the gulf of Mexico. Asteroids hurtle round the sun in a belt which lies inside the orbit of Jupiter. There are plenty of asteroids out there – small ones are hitting us all the time – and a few of them are large enough to cause cataclysmic extinctions if they were to hit us. The comets have larger, eccentric orbits around the sun, mostly well outside what we conventionally think of as the solar system, but occasionally coming inside it, as Halley's comet does every 76 years and the Hale Bopp comet every 4,000 years or so. Perhaps the Permian event was caused by an even larger comet strike than the Cretaceous one. Perhaps the suggested 26-million-year cycle of mass extinctions is caused by a rhythmic boost in the rate of comet strikes.

But why should comets become more likely to hit us every 26

million years? Here we launch ourselves into deep speculation. It has been suggested that the sun has a sister star, and the two orbit each other with a periodicity of about 26 million years. This hypothetical binary partner, which has never been seen but which has nevertheless been given the dramatic name Nemesis, passes, once per orbital rotation, through the so-called Oort Cloud, the belt of perhaps a trillion comets which orbits the sun beyond the planets. If there was a Nemesis that passed close to, or through, the Oort Cloud, it is plausible that it would disturb the comets, and this might increase the likelihood of one of them hitting earth. If this all happened – and the chain of reasoning is admittedly tenuous – it could account for the 26-million-year periodicity of mass extinctions that some people think the fossil record shows. It is a pleasing thought that mathematical unweaving of the noisy spectrum of animal extinctions might be the only means we have of detecting an otherwise unknown star.

Starting with the ultra high frequencies of light and other electromagnetic waves, we passed, via the intermediate frequencies of sound and the swinging elephant's penis, to ultra low frequencies and the alleged 26-million-year wavelength of mass extinctions. Let's return to sound, and in particular that crowning feat of the human brain, the weaving and unweaving of speech sounds. The vocal 'cords' are really a pair of membranes which vibrate together in the breathing passage like a pair of woodwind reeds. Consonants are produced as more or less explosive interruptions of the air flow, caused by closure and contact of the lips, teeth, tongue and back of throat. Vowels vary in the same kind of way as trumpets differ from oboes. We make different vowel sounds rather as a trumpeter moves a mute in and out, to shift the preponderant sine waves summing into the composite sound. Different vowels have different combinations of harmonics above the fundamental frequency. The fundamental frequency itself, of course, is lower for men than for women and children, yet male vowels sound similar to the corresponding female vowels because of the pattern of harmonics.

Each vowel sound has its own characteristic pattern of frequency stripes, like barcodes once again. In the study of speech, the barcode stripes are called 'formants'.

Any one language, or dialect within a language, has a finite list of vowel sounds, and each of those vowel sounds has its own formant barcode. Other languages, and different accents within languages, have different vowel sounds which are made by holding the mouth and tongue in intermediate positions, again as a trumpeter disposes the mute in the bell of the instrument. Theoretically there is a continuous spectrum of vowel sounds. Any one language employs a useful selection, a discontinuous repertoire picked out from the continuous spectrum of available vowels. Different languages pick out different points along the spectrum. The vowel in the French *tu* and the German *über*, which doesn't occur in (my version of) English, is approximately intermediate between *oo* and *ee*. It doesn't too much matter which landmark points along the spectrum of available vowels a language picks on, so long as they are spaced far enough apart to avoid ambiguity within that language.

The story for consonants is more complicated, but there is a similar range of consonant barcodes, with actual languages employing a limited subset from those available. Some languages employ sounds which are far off the spectrum of the majority of languages, for example the clicks of some southern African tongues. As with vowels, different languages parcel up the available repertoire differently. Several of the languages of the Indian subcontinent have a dental sound which is intermediate between the English 'd' and 't'. The French hard 'c' as in *comme* is intermediate between the English hard 'c' and hard 'g' (and the 'o' is intermediate between the English vowels in cod and cud). The tongue, lips and voice can be modulated to produce an almost infinite variety of consonants and vowels. When the barcodes are patterned in time to form phonemes, syllables, words and sentences, the range of ideas that can be communicated is unlimited.

Stranger yet, the things that can be communicated include

images, ideas, feelings, love and exultation – the kind of thing that Keats does so sublimely.

> *My heart aches, and a drowsy numbness pains*
> *My sense, as though of hemlock I had drunk,*
> *Or emptied some dull opiate to the drains*
> *One minute past, and Lethe-wards had sunk:*
> *'Tis not through envy of thy happy lot,*
> *But being too happy in thy happiness –*
> *That thou, light-wingèd Dryad of the trees,*
> *In some melodious plot*
> *Of beechen green, and shadows numberless,*
> *Singest of summer in full-throated ease.*
>
> 'Ode to a Nightingale' (1820)

Read the words aloud and the images tumble into your brain, as if you really were drugged by a nightingale's song in a leafy summer beechwood. At one level it is all done by a pattern of air pressure waves, a pattern whose richness is first unwoven into sine waves in the ear and then rewoven together in the brain to reconstruct images and emotions. Stranger yet, the pattern can be broken down mathematically into a stream of numbers, and it retains its power to transport and haunt the imagination. When a laser disc (CD) is made, say, of the *Saint Matthew Passion*, the rising and falling pressure wave, with all its wiggles and kinks, is sampled at frequent intervals and translated into digital data. The digits could, in principle, be printed as dull, black and white zeroes and ones on reams of paper. Yet the numbers retain the power, if transduced back into pressure waves, to move a listener to tears.

Keats may not have intended it literally, but the idea of nightingale song working as a drug is not totally far-fetched. Consider what it is doing in nature, and what natural selection has shaped it to do. Male nightingales need to influence the behaviour of female nightingales, and of other males. Some ornithologists have thought of song as conveying information: 'I am a male of the species *Luscinia megarhynchos*, in breeding condition, with a territory, hormonally primed to mate and build a nest.' Yes, the song does contain that information, in the sense that a female who acts on the assumption that it is true could benefit thereby. But another way to look at it has always seemed to me more vivid. The song is not informing the female but *manipulating* her. It is not so much changing what the female knows as directly changing the internal physiological state of her brain. It is acting like a drug.

There is experimental evidence from measuring the hormone levels of female doves and canaries, as well as their behaviour, that the sexual state of females is directly influenced by the vocalizations of males, the effects being integrated over a period of days. The sounds from a male canary flood through the female's ears into her brain where they have an effect that is indistinguishable from one that an experimenter can procure with a hypodermic syringe. The male's 'drug' enters the female through the portals of her ears rather than through a hypodermic, but this difference does not seem particularly telling.

The idea that birdsong is an auditory drug gains plausibility when you look at how it develops during the individual's lifetime. Typically, a young male songbird teaches himself to sing by practising: matching up fragments of trial song against a 'template' in his brain, a pre-programmed notion of what the song of his species 'ought' to sound like. In some species, such as the American song sparrow, the template is built in, programmed by the genes. In other species, such as the white crowned sparrow or the European chaffinch, it is derived from a 'recording' of another male's song, made early in the young male's life from listening to an adult. Wherever the template comes from, the young male teaches himself how to sing in such a way as to match it.

That, at least, is one way to talk about what happens when a young bird perfects his song. But think of it another way. The song is ultimately designed to have a strong effect on the nervous system of another member of the species, either a prospective mate or a possible territorial rival who needs to be warned off. But the young bird himself is a member of his own species. His brain is a typical brain from that species. A sound that is effective in arousing his own emotions is likely to be as effective in arousing a female of the same species. Instead of speaking of the young male trying to shape his practice song to 'match' a built-in 'template', we could think of him as practising on himself as a typical member of his species, trying out fragments of song to see whether they excite his own passions, that is, experimenting with his own drugs on himself.

And, to complete the circuit, perhaps it is not too surprising that nightingale song should have acted like a drug on the nervous system of John Keats. He was not a nightingale, but he was a vertebrate, and most drugs that work on humans have a comparable effect upon other vertebrates. Manmade drugs are the products of comparatively crude trial-and-error testing by chemists in the laboratory. Natural selection has had thousands of generations in which to fine-tune its drug technology.

Should we feel indignant on Keats's behalf at such a comparison? I do not believe that Keats himself would have done so – Coleridge even less. The 'Ode to a Nightingale' accepts the implication of the drug analogy, makes it wonderfully real. It is not demeaning to human emotion that we try to analyse and explain it, any more than, to a balanced judge, the rainbow is diminished when a prism unweaves it.

In this chapter and the previous one, I have used the barcode as a symbol of precise analysis, in all its beauty. Mixed light is sorted into its rainbow of component colours and everybody sees beauty. That is a first analysis. Closer detail reveals fine lines and a new elegance, the elegance of detection, of the bringing of order and understanding. Fraunhofer barcodes speak to us of the exact elemental nature of distant stars. A precisely measured pattern of stripes is a coded message from across the parsecs. There is grace in the sheer *economy* of unweaving intimate details about a star which, one had thought, could be found only through the costly undertaking of a journey lasting 2,000 human lifetimes. On another scale, we find a similar story when we look at the formant stripes in speech, the harmonic barcodes of music. There is elegance, too, in the barcodes of dendrochronology: the stripes across ancient *Sequoia* wood which tell us precisely in which year BC the tree was seeded, and what the weather was like in every one of the intervening years (for weather conditions are what give tree rings their characteristic widths). Like Fraunhofer's lines transmitted across space, tree rings transmit messages to us across time, and again there is a supple economy. It is the power – the fact that we can learn so much by precise analysis of what seems so little information – that gives these unweavings their beauty. The same is true, perhaps even more dramatically, of sound waves in speech and music – barcodes on the air.

Recently we have been hearing much about another kind of barcode – DNA 'fingerprints', barcodes in the blood. DNA barcodes expose and reconstruct details of human affairs that one might have supposed forever inaccessible even to legendarily great detectives. The main practical use of barcodes in the blood so far is in courts of law, and it is to them – and the benefits that a scientific attitude may bring to them – that we turn in the next chapter.

5
BARCODES AT THE BAR

And he said, Woe unto you also, ye lawyers! for ye lade men with burdens grievous to be borne, and ye yourselves touch not the burdens with one of your fingers. . . . Woe unto you, lawyers! for ye have taken away the key of knowledge: ye entered not in yourselves, and them that were entering in ye hindered. *Luke 11*

On the face of it, the law may seem about as far as you can get from poetry or the wonder of science. Perhaps there is poetic beauty in the abstract ideas of justice or fairness, but I doubt if many lawyers are moved by it. In any case, that is not what this chapter is about. I shall be looking at an example of the role of science in the law: at a different aspect of science and its importance in society; a sense in which scientific understanding may become a valuable part of good citizenship. In courts of law, juries are increasingly asked to understand evidence which the lawyers themselves may not fully comprehend. Evidence from the unweaving of DNA – what we shall come to see as barcodes in the blood – is the outstanding example, and it is the main subject of this chapter. But it is not just facts about DNA that scientists can contribute. More importantly, it is the underlying theory of probability and statistics; it is scientific ways of making inferences that need to be brought to bear. Such matters stretch beyond the narrow subject of DNA evidence.

I am told on good authority that defence lawyers in the United States sometimes object to jury candidates on the grounds that they have had a scientific education. What can this mean? I would not question the right of defence lawyers to disallow the selection of particular jurors. A juror may be prejudiced against the race or class to which the defendant belongs. It is obviously undesirable that a raving homophobe should try a case of anti-homosexual violence. It is for this kind of reason that defence lawyers in some countries are allowed to cross-examine potential jurors and strike them off the list. In the USA lawyers can be completely blatant about their criteria for jury selection. A colleague tells me of a time when he was up for selection to a jury, on an injury litigation case. The lawyer asked, 'Would anyone here have a problem awarding a *substantial* amount of money to my client, perhaps in the millions?'

A lawyer can also disqualify a juror without giving reasons. Although this may be just, the only time I have seen it happen it misfired. I was a member of a panel of 24 individuals from which juries of 12 were to be selected. I had already participated in two juries with members of this panel, and I knew their individual foibles. One particular man was cast-iron prosecution fodder; he would take the same hard line almost regardless of the particular case. The defence lawyer waved him through like a breeze. The next one up, a large middle-aged woman, was the opposite: a guaranteed softie, a pure gift to the defence. But her appearance perhaps suggested the opposite, and it was against her that the defence lawyer chose to exercise his right of veto. I have never forgotten the look of wounded hurt on her face as, with a cutting movement of the hand, learned counsel struck her – whom he little knew could have been his secret weapon – out of the jury box.

But, to repeat the astonishing fact, lawyers in the United States have been known to use the following reason for striking down potential jurors: the prospective juror is well educated in science, or has some knowledge of genetics or probability theory. What is the problem? Are geneticists known to harbour deep-seated prejudices against certain sections of society? Are mathematicians especially likely to be of the 'flog 'em . . . string 'em up . . . it's the only language they understand . . . law and order' persuasion? Of course not. Nobody has ever claimed such a thing.

The lawyers' objections are more ignobly based. There is a new kind of evidence increasingly coming into the criminal courts: evidence from DNA fingerprinting, and it is extremely powerful. If your client is innocent, DNA evidence may well provide a knock-down convincing way to establish his innocence. Conversely, if he is guilty, DNA evidence has a good chance of establishing his guilt in cases where no other evidence can. DNA evidence is quite hard to understand at the best of times. There are controversial aspects of it which are even harder. In these circumstances, you would think that an honest lawyer who wishes to see justice done would welcome jurors capable of grasping the arguments. Wouldn't it be an obviously good thing to have at least one or two people in the jury room who can redress the ignorance of their baffled colleagues? What kind of a lawyer is it who prefers a jury incapable of following the case that either attorney is making?

The answer is a lawyer who is more interested in winning than in seeing justice done. A lawyer, in other words. And it seems to be a fact that advocates, of both prosecution and defence, frequently disallow individual jurors specifically because they are educated in science.

Courts of law have always needed to establish individual identity. Was the individual seen hurrying from the scene Richard Dawkins? Is the hat dropped at the scene of the crime his hat? Are those his fingerprints on the weapon? A yes answer to one of these questions does not by itself prove his guilt, but it is certainly an important factor to be taken into account. Most of us, including most jurors and lawyers, have an intuitive sense that there is something specially reliable about eye-witness evidence. In this we are almost certainly wrong, but the error is a pardonable one. It may even be built into us by millennia of evolutionary history in which eye-witness evidence really was the most reliable. If I see a man in a red woolly hat climbing a drainpipe, you will have a hard time persuading me later that he was actually wearing a blue beret. Our intuitive biases are such that eye-witness evidence trumps all other categories. Yet numerous studies have shown that eye-witnesses, however convinced they may be, however sincere and well-meaning, frequently misremember even conspicuous details such as the colour of clothing and the number of assailants present.

When individual identification is important, for instance when a woman who has been raped is called upon to identify her attacker, courts perform a rudimentary statistical test known as the identity parade or line-up. The woman is led past a line of men, one of whom the police suspect on other grounds. The others have been pulled in off the streets or are out-of-work actors, or police officers dressed in plain clothes. If the woman picks out one of these stooges, her identification evidence is discounted. But if she picks out the man the police already suspect, her evidence is taken seriously.

Rightly so. Especially if the number of people in the identity parade is large. We are all statisticians enough to see why this is. The prior suspicion of the police must be open to doubt – otherwise there would be no point in seeking the woman's evidence at all. What impresses us is agreement between the woman's identification and the independent evidence offered by the police. If the identity parade contains only two men, the witness would have a 50 per cent chance of picking the man already suspected by the police, even if she chose at random – or if she were mistaken. Since the police might also be mistaken, this represents an unacceptably high risk of injustice. But if there are 20 men in the line, the woman has only a 1 in 20 chance of choosing, by guesswork or error, the man the police already suspect. The coincidence of her identification and the police's prior suspicion probably really means something. What is going on here is the assessment of coincidence, or the odds that something might happen by chance alone. The probability of meaningless coincidence is even less if the identity parade has 100 men, because a 1 in 100 chance of error is noticeably less than a

1 in 20 chance of error. The longer the line-up, the more secure the eventual conviction.

We also have an intuitive sense that the men chosen for the line-up must not look too obviously different from the suspect. If the woman originally told the police to look for a man with a beard, and the police have now arrested a bearded suspect, it is clearly unjust to stand him in a line with 19 clean-shaven men. He might as well be standing by himself. Even if the woman has said nothing about the appearance of her attacker, if the police have arrested a punk in a leather jacket it would be wrong to stand him in a line of suited accountants with furled umbrellas. In multiracial countries such considerations have added importance. Everyone understands that a black suspect should not be placed in an otherwise all-white line-up, or vice versa.

When we think about how we identify somebody, the face first leaps to mind. We are particularly good at distinguishing faces. As we shall see in another connection, we even seem to have evolved a special part of the brain set aside for the purpose, and certain kinds of brain damage disable our face-recognition faculty while leaving the rest of vision intact. In any case, faces are good for recognition because they are so variable. With the well-known exception of identical twins, you seldom meet two people whose faces are confusable. It is not totally unknown, however, and an actor can be made up to look very like somebody else. Dictators often employ doubles to perform for them when they are too busy, or to draw the fire of assassins. It has been suggested that one reason charismatic leaders so often sport moustaches (Hitler, Stalin, Franco, Saddam Hussein, Oswald Mosley) is to make it easier for doubles to impersonate them. Mussolini's shaven head perhaps served the same purpose.

Apart from identical twins, ordinary close relatives are sometimes sufficiently alike to fool people who don't know them well. (Unfortunately the story that Doctor Spooner, when Warden of my college, once stopped an undergraduate and said, 'I never can remember, is it you or your brother was killed in the war?' is probably not true, like most alleged Spoonerisms.) The resemblance of brothers and sisters, of fathers and sons, of grandparents and grandchildren, serves to remind us of the huge pool of facial variety in the general population of non-relatives.

But faces are only a special case. We are riddled with idiosyncrasies which, with sufficient training, can be used to identify individuals. I had a schoolfriend who claimed (and my spot checks confirmed it) that he could recognize any member of the 80-strong residence in which we lived purely by listening to their footsteps. I had another friend from Switzerland who claimed that when she walked into a room she could tell, by smell, which members of her circle of acquaintances had recently left the room. It is not that her colleagues didn't wash, just that she was unusually sensitive. That this is in principle possible is confirmed by the fact that police dogs can distinguish between any two human beings by smell alone, with the exception, yet again, of identical twins. As far as I know, the police haven't adopted the following technique, but I bet you could train bloodhounds to track down a kidnapped child after giving them a sample sniff of his brother. A way might even be found to use a jury of bloodhounds to decide paternity cases.

Voices are as idiosyncratic as faces, and various research teams are working on computer voice recognition systems for authenticating identity. It would be a great boon if, in the future, we could dispense with front door keys and rely on a voice-operated computer to obey our personal Open Sesame command. Handwriting is sufficiently individual for the written signature to be used as a guarantee of identity on bank cheques and important legal documents. Signatures are actually not particularly secure because they are too easily forged, but it is still impressive how recognizable handwriting can be. A promising newcomer to the list of individual 'signatures' is the iris of the eye. At least one bank is experimenting with automated iris-scanning machines as a way of verifying identity. The customer stands in front of a camera which photographs the eye, digitizes the image into what a newspaper described as 'a 256 byte human bar code'. But none of these methods of verifying human identity even comes close to the potential of DNA fingerprinting, properly applied.

It is not surprising that police dogs can smell the difference between any two humans except identical twins. Our sweat contains a complicated cocktail of proteins, and the precise details of all proteins are minutely specified by the coded DNA instructions that are our genes. Unlike handwriting and faces, which vary continuously and grade smoothly into one another, genes are digital codes, much like those used in computers. Again with the exception of identical twins, we differ genetically from all other people in discrete, discontinuous ways: an exact number of ways that you could even count if you had the patience. The DNA in each one of my cells (give or take a tiny minority of mistakes, and not including red blood cells which have lost all their DNA, or reproductive cells which contain a random half of my genes) is identical to the DNA in all my other cells. It differs from the DNA in every one of your cells, not in some vague, impressionistic way but at a precise number of locations dotted along the billions of DNA letters that we both have.

It is almost impossible to exaggerate the importance of the digital revolution in molecular genetics. Before Watson and Crick's epochal announcement in 1953 of the structure of DNA, it was still possible to agree with the concluding words of Charles Singer's authoritative *A Short History of Biology*, published in 1931:

... despite interpretations to the contrary, the theory of the gene is not a 'mechanist' theory. The gene is no more comprehensible as a chemical or physical entity than is the cell or, for that matter, the organism itself. Further, though the theory speaks in terms of genes as the atomic theory speaks in terms of atoms, it must be remembered that there is a fundamental distinction between the two theories. Atoms exist independently, and their properties as such can be

examined. They can even be isolated. Though we cannot see them, we can deal with them under various conditions and in various combinations. We can deal with them individually. Not so the gene. It exists only as a part of the chromosome, and the chromosome only as part of a cell. If I ask for a living chromosome, that is, for the only effective kind of chromosome, no one can give it to me except in its living surroundings any more than he can give me a living arm or leg. The doctrine of the relativity of functions is as true for the gene as it is for any of the organs of the body. They exist and function only in relation to other organs. Thus the last of the biological theories leaves us where the first started, in the presence of a power called life or psyche which is not only of its own kind but unique in each and all of its exhibitions.

This is dramatically, profoundly, hugely wrong. And it really matters. Following Watson and Crick and the revolution that they sparked, a gene *can* be isolated. It can be purified, bottled, crystallized, read as digitally coded information, printed on a page, fed into a computer, read out again into a test tube and reinserted into an organism where it works exactly as it did before. When the Human Genome Project, which set out to work out the complete gene sequence of a human being, is completed, probably by the year 2003, the full genome will fit comfortably on two standard CD ROM discs, leaving enough space for a textbook of molecular embryology. These two discs could then be sent into outer space, and the human race could go extinct secure in the knowledge that there is now a chance that at some future time and in some distant place, a sufficiently advanced civilization would be able to reconstitute a human being. Meanwhile, back on earth, it is because DNA is deeply and fundamentally digital – because the differences between individuals and between species can be precisely counted, not vaguely and impressionistically measured – that DNA fingerprinting is potentially so powerful.

I assert the uniqueness of each individual's DNA with confidence, but even this is only a statistical judgement. Theoretically, the sexual lottery could throw up the same genetic sequence twice. An 'identical twin' of Isaac Newton could be born tomorrow. But the number of people that would have to be born in order to make this event at all likely would be larger than the number of atoms in the universe.

Unlike our face, voice or handwriting, the DNA in most of our cells stays the same from babyhood to old age, and it cannot be altered by training or cosmetic surgery. Our DNA text has such a huge number of letters that we can precisely quantify the expected number shared by, say, brothers or first cousins as opposed to, say, second cousins or random pairs chosen from the population at large. This makes it useful not only for labelling individuals uniquely and matching them to traces such as blood or semen, but for establishing paternity and other genetic relationships. British law allows people to immigrate if they can prove that their parents are already British citizens. A number of children from the Indian subcontinent have been arrested by sceptical immigration officials. Before the advent of DNA fingerprinting it was often impossible for these unfortunate people to prove their parentage. Now it is easy. All you do is take a sample of blood from the putative parents and compare a particular set of genes with the corresponding set of genes from the child. The verdict is clear and unequivocal, with none of the doubt or fuzziness that creates a need for qualitative judgements. Several young people in Britain today owe their citizenship to DNA technology.

A similar method was used to identify skeletons discovered in Yekaterinburg and suspected of belonging to the executed Russian royal family. Prince Philip, Duke of Edinburgh, whose exact relationship to the Romanovs is known, graciously gave blood, and from this it was possible to establish that the skeletons were indeed those of the Tsar's family. In a more macabre case, a skeleton exhumed in South America was proved to belong to Doctor Josef Mengele, the Nazi war criminal known as the 'Angel of Death'. DNA taken from the bones was compared with blood from Mengele's still-living son, and the identity of the skeleton proved. More recently, a corpse dug up in Berlin has been proved, by the same method, to be that of Martin Bormann, Hitler's deputy, whose disappearance had led to endless legends and rumours and more than 6,000 'sightings' around the world.

Despite the name 'fingerprinting', our DNA, being digital, is even more individually characteristic than the patterns of whorls on our fingers. The name is appropriate because, like true fingerprints, DNA evidence is often inadvertently left behind after a person has departed the scene. DNA can be extracted from a bloodstain on a carpet, from semen inside a rape victim, from a crust of dried nasal mucus on a handkerchief, from sweat or from shed hairs. The DNA in the sample can then be compared with that in the blood taken from a suspect. It is possible to assess, to almost any desired level of probability, whether the sample belongs to a particular person or not.

So, what are the snags? Why is DNA evidence controversial? What is it about this important kind of evidence that makes it possible for lawyers to bamboozle juries into misinterpreting or ignoring it? Why have some courts been moved to the despairing extreme of ruling out this evidence altogether?

There are three major classes of potential problem, one simple, one sophisticated and one silly. I'll come to the silly problem and the more sophisticated difficulties later but first, as with any kind of evidence, there is the simple – and very important – possibility of human error. Possibilities, rather, for there are plenty of opportunities for mistakes and even sabotage. A tube of blood may be mislabelled, either by accident or in a deliberate attempt to frame somebody. A sample from the scene of a crime may be contaminated by sweat from a lab technician or a police officer. The danger of contamination is especially great in those cases where an ingenious technique of amplification called PCR (polymerase chain reaction) is used.

You can easily see why amplification might be desirable. A tiny smear of sweat on a gun butt contains precious little DNA. Sensitive though DNA analysis can be, it needs a certain minimum quantity of material to work on. The technique of PCR, invented in 1983 by the American biochemist Kary B. Mullis, is the dramatically successful answer. PCR takes what little DNA there is and produces millions of copies, multiplying again and again whatever code sequences are there. But, as always with amplification, errors are amplified along with the true signal. Stray scraps of DNA contamination from a technician's sweat are amplified as effectively as the specimen from the scene of the crime, with obvious possibilities for injustice.

But human error is not peculiar to DNA evidence. All kinds of evidence are vulnerable to bungling and sabotage, and must be handled with scrupulous care. The files in a conventional fingerprint library may be mislabelled. The murder weapon may have been touched by innocent people as well as the murderer, and their fingerprints have to be taken, along with the suspect's, for elimination purposes. Courts of law are already accustomed to the need to take all possible precautions against mistakes and they still, sometimes tragically, happen. DNA evidence is not immune to human bungling but nor is it particularly vulnerable, except in so far as PCR amplifies error. If all DNA evidence were to be thrown out because of occasional mistakes, the precedent should rule out most other kinds of evidence, too. We have to suppose that codes of practice and rigorous precautions can be developed to guard against human error in the presentation of all kinds of legal evidence.

The more sophisticated difficulties that bedevil DNA evidence will take longer to explain. They, too, have their precedents in conventional types of evidence, although this point often does not seem to be understood in law courts.

Where identification evidence of any kind is concerned, there are two types of error which correspond to the two types of error in any statistical evidence. In another chapter, we shall call them

Type 1 and Type 2 errors, but it is easier to think of them as false positive and false negative. A guilty suspect may escape, through not being recognized – false negative. And – false positive (which most people would see as the more dangerous error) – an innocent suspect may be convicted because he happens, by ill luck, to resemble the genuinely guilty party. In the case of ordinary eye-witness identification, an innocent bystander who happens to look a bit like the real criminal could consequently be arrested – false positive. Identity parades are designed to make this less probable. The chance of a miscarriage of justice is inversely related to the number of people standing in the line-up. The danger can be increased in the ways we have already considered – the line-up being unfairly stacked with clean-shaven men for example.

In the case of DNA evidence the danger of a false positive conviction is theoretically very low indeed. We have a blood sample from a suspect, and we have a specimen from the scene of the crime. If the entire set of genes in both these samples could be written down, the probability of a false conviction is one in billions and billions. Identical twins apart, the chance that any two humans would match all their DNA is tantamount to zero. But unfortunately it is not practical to work out the complete gene sequence of a human being. Even after the Human Genome Project is completed, to attempt the equivalent in the solution of each crime is unrealistic. In practice, forensic detectives concentrate on small sections of the genome, preferably sections that are known to vary in the population. And now our fear must be that, although we could safely rule out misidentification if the whole genome were considered, there might be a danger of two individuals' being identical with respect to the small portion of DNA that we have time to analyse.

The probability that this would happen ought to be measurable for any particular section of the genome; we could then decide whether it was an acceptable risk. The larger the section of DNA, the smaller the probability of error, just as, in an identity parade, the longer the line-up the safer the conviction. The difference is that an identity parade, in order to compete with the DNA equivalent, would need to contain not a couple of dozen people but thousands, millions or even billions in the line. Apart from this quantitative difference, the analogy with the identity parade continues. We shall see that there is a DNA equivalent of our hypothetical line-up of clean-shaven men with one bearded suspect. But first, a little more background on DNA fingerprinting.

Obviously we sample the equivalent parts of the genome in both suspect and specimen. These parts of the genome are chosen for their tendency to vary widely in the population. A Darwinian would note that the parts that don't vary are often the parts that have an important role to play in the survival of the organism. Any substantial variations in these important genes are likely to have been removed from the population by the death of their possessors – Darwinian natural selection. But there are other parts of the genome that are very variable, perhaps because they are not important for survival. This isn't the whole story because in fact some useful genes are quite variable. The reasons for this are controversial. It's a bit of a digression but . . . What is this life if, full of stress, we have no freedom to digress?

The 'neutralist' school of thought, associated with the distinguished Japanese geneticist Motoo Kimura, believes that useful genes are *equally* useful in a variety of different forms. This emphatically does not mean that they are useless, only that the different forms are equally good at what they do. If you think of genes as writing out their recipes in words, the alternative forms of a gene can be thought of as the very same words written in different typefaces: the meaning is the same, and the product of the recipe will come out the same. Genetic changes, 'mutations', that make no difference are not 'seen' by natural selection. They aren't mutations at all, for all the difference they make to the life of the animal, but they are potentially useful mutations from the point of view of the forensic scientist. The population ends up with lots of variety at such a locus (position in a chromosome), and this kind of variety could in principle be used for fingerprinting.

The other theory of variation, opposed to Kimura's neutral theory, believes that the different versions of the genes really do different things and that there is some special reason why both are preserved by natural selection in the population. For example, there might be two alternative forms of a blood protein, α and β, which are susceptible to two infectious diseases called alfluenza and betaccosis respectively, each being immune to the other disease. Typically, an infectious disease needs a critical density of susceptible victims in a population, otherwise an epidemic can't get going. In a population dominated by α types, there are frequent epidemics of alfluenza but not of betaccosis. So natural selection favours the β types who are immune to alfluenza. It favours them so much that after a while they come to dominate the population. Now the tables are turned. There are epidemics of betaccosis, but not of alfluenza. The α types now are favoured by natural selection because they are immune to betaccosis. The population may keep oscillating between α dominance and β dominance, or it may settle down to an intermediate mixture, an 'equilibrium'. Either way, we'll see plenty of variation at the gene locus concerned, and this is good news for the fingerprinters. The phenomenon is called 'frequency dependent selection' and it is one suggested reason for high levels of genetic variation in the population. There are others.

However, for our forensic purposes, it matters only that there are variable sections of the genome. Whatever the verdict in the controversy over whether the useful bits of the genome are variable, there are in any case lots of other regions of the genome which are never even read, or never translated into their protein equivalents. Indeed, an astonishingly high proportion of our genes seem to be doing nothing whatsoever. They are therefore free to vary, which makes them excellent DNA fingerprinting material.

As if to confirm the fact that a great deal of DNA is doing nothing useful, the sheer quantity of DNA in the cells of different kinds of organisms is wildly variable. Since DNA information is digital, we can measure it in the same kind of units as we measure computer information. One bit of information is enough to specify one yes/no decision: a 1 or a 0, a *true* or a *false*. The computer on which I am writing this has 256 megabits (32 megabytes) of core memory. (The first computer that I owned was a bigger box but had less than one five thousandth of the memory capacity.) The equivalent fundamental unit in DNA is the nucleotide base. Since there are 4 possible bases, the information content of each base is equivalent to 2 bits. The common gut bacterium *Escherichia coli* has a genome of 4 megabases or 8 megabits. The crested newt, *Triturus cristatus*, has 40,000 megabits. The 5,000-fold ratio between crested newt and bacterium is about the same as that between my present computer and my first one. We humans have 3,000 megabases or 6,000 megabits. This is 750 times as great as the bacterium (which satisfies our vanity), but what are we to make of the newt trumping us sixfold? We'd like to think that genome size is not strictly proportional to what it does: presumably quite a lot of that newt DNA isn't doing anything. This is certainly true. It is also true of most of our DNA. We know from other evidence that, of the 3,000 megabase human genome, only about 2 per cent is actually used for coding protein synthesis. The rest is often called junk DNA. Presumably the crested newt has an even higher percentage of junk DNA. Other newts have not.

The surplus of unused DNA falls into various categories. Some of it looks like real genetic information, and probably represents old, defunct genes, or out-of-date copies of genes that are still in use. These pseudo-genes would make sense if they were read and translated. But they are not read and translated. Hard disks on computers usually contain comparable junk: old copies of work in progress, scratchpad space used by the computer for interim operations, and so on. We users don't see this junk, because our computers only show us those parts of the disk that we need to know about. But if you get right down and read the actual information on the disk, byte by byte, you'll see the junk, and much of it will make some sort of sense. There are probably dozens of disjointed fragments of this very chapter peppered around my hard disk at present, although there is only one 'official' copy that the computer tells me about (plus a prudent back-up).

In addition to the junk DNA which *could* be read but isn't, there is plenty of junk DNA which not only isn't read but wouldn't make any sense if it were. There are huge stretches of repeated nonsense, perhaps repeats of one base, or alternations of the same two bases, or repeats of a more complicated pattern. Unlike the other class of junk DNA, we cannot account for these 'tandem repeats' as outdated copies of useful genes. This repetitive DNA has never been decoded, and presumably has never been of any use. (Never useful for the animal's survival, anyway. From the point of view of the selfish gene, as I explained in another book, we could say that any kind of junk DNA is 'useful' to itself if it just keeps surviving and making more copies of itself. This suggestion has come to be known by the catchphrase 'selfish DNA', although this is a little unfortunate because, in my original sense, working DNA is selfish too. For this reason, some people have taken to calling it 'ultraselfish DNA'.)

Anyway, whatever the reason, junk DNA is there, and there in prodigious quantities. Because it is not used, it is free to vary. Useful genes, as we have seen, are severely constrained in their freedom to change. Most changes (mutations) make a gene work less effectively, the animal dies and the change is not passed on. This is what Darwinian natural selection is all about. But mutations in junk DNA (mostly changes in the number of repeats in a given region) are not noticed by natural selection. So, as we look around the population, we find most of the variation that is useful for fingerprinting in the junk regions. As we shall now see, tandem repeats are particularly useful because they vary with respect to number of repeats, a gross feature which is easy to measure.

If it wasn't for this, the forensic geneticist would need to look at the exact sequence of bases in our sample region. This can be done, but sequencing DNA is time-consuming. The tandem repeats allow us to use cunning short-cuts, as discovered by Alec Jeffreys of the University of Leicester, rightly regarded as the father of DNA fingerprinting (and now Sir Alec). Different people have different numbers of tandem repeats in particular places. I might have 147 repeats of a particular piece of nonsense, where you have 84 repeats of the same piece of nonsense in the corresponding place in your genome. In another region, I might have 24 repeats of a particular piece of nonsense to your 38 repeats. Each of us has a characteristic fingerprint consisting of a set of numbers. Each of these numbers in our fingerprint is the number of times a particular piece of nonsense is repeated in our genome.

We get our tandem repeats from our parents. We each have 46 chromosomes, 23 from our father and 23 homologous, or corresponding, chromosomes from our mother. These chromosomes come complete with tandem repeats. Your father got his 46 chromosomes from your paternal grandparents, but he didn't pass them on to you in their entirety. Each of his mother's chromosomes was lined up with its paternal opposite number and bits were exchanged before a composite chromosome was put into the sperm that helped to make you. Every sperm and every egg is unique because it is a different mix of maternal and paternal chromosomes. The mixing process affects the tandem repeat sections as well as the meaningful sections of the chromosomes. So our characteristic numbers of tandem repeats are inherited, in much the same way as our eye colour and hair curliness are inherited. With the difference that, whereas our eye colour results from some kind of joint verdict of our paternal and our maternal genes, our tandem repeat numbers are properties of the chromosomes themselves and can therefore be measured separately for paternal and maternal chromosomes. At any particular tandem repeat region, each of us has two readings: a paternal chromosome repeat number and a maternal chromosome repeat number. From time to time, chromosomes mutate – suffer a random change – in their tandem repeat numbers. Or a particular tandem region may be split by chromosomal crossing over. This is why there is variation in tandem repeat numbers in the population. The beauty of tandem repeat numbers is that they are easy to measure. You don't have to get embroiled in detailed sequencing of coded DNA bases. You do something a bit like weighing them. Or, to take another equally apt analogy, you spread them out like coloured bands from a prism. I'll explain one way of doing this.

First you need to make some preparations. You make a so-called DNA probe, which is a short sequence of DNA that exactly matches the nonsense sequence in question – up to about 20 nucleotide bases long. This is not difficult to do nowadays. There are several methods. You can even buy a machine off the shelf which makes short DNA sequences to any specification, just as you can buy a keyboard to punch any desired string of letters on a paper tape. By supplying the synthesizing machine with radioactive raw materials, you make the probes themselves radioactive, and so 'label' them. This makes the probes easy to find again later, as natural DNA is not radioactive, and so the two are readily distinguishable from each other.

Radioactive probes are a tool of the trade, which you must have ready before you start a Jeffreys fingerprinting exercise. Another essential tool is the 'restriction enzyme'. Restriction enzymes are chemical tools that specialize in cutting DNA, but cutting it only in particular places. For example, one restriction enzyme may search the length of a chromosome until it finds the sequence GAATTC (G, C, T and A are the four letters of the DNA alphabet; all genes, from all species on earth, differ only in consisting of different sequences of these four letters). Another restriction enzyme cuts the DNA wherever it can find the sequence GCGGCCGC. A number of different restriction enzymes are available in the toolbox of the molecular biologist. They originate from bacteria, who use them for their own defensive purposes. Each restriction enzyme has its own unique search string which it homes in on and cuts.

Now, the trick is to choose a restriction enzyme whose specific search string is completely absent from the tandem repeat we are interested in. The whole length of DNA is therefore chopped into short stretches, bounded by the characteristic search string of the restriction enzyme. Of course, not all the stretches will consist of the tandem repeat we are looking for. All sorts of other stretches of DNA will happen to be bounded by the favoured search string of the restriction enzyme scissors. But some of them will consist of tandem repeats and the length of each scissored stretch will be largely determined by the number of tandem repeats in it. If I have 147 repeats of a particular piece of DNA nonsense, where you have only 83, my snipped fragments will be correspondingly longer than your snipped fragments.

We can measure these characteristic lengths using a technique that has been around in molecular biology for quite a while. This is the bit that is rather like spreading them out with a prism, as Newton did for white light. The standard DNA 'prism' is a gel electrophoresis column, that is, a long tube filled with jelly through which an electric current is passed. A solution containing the scissored stretches of DNA, all jumbled together, is poured into one end of the tube. The DNA fragments are all electrically attracted to the negative end of the column, which is at the other end of the tube, and they move steadily through the jelly. But they don't all move at the same rate. Like light of low vibration frequency moving through glass, small fragments of DNA move faster than large ones. The result is that, if you switch the current off after a suitable interval, the fragments have spread themselves out along the column, just as Newton's colours spread themselves out because light from the blue end of the spectrum is more readily slowed down by glass than light from the red end.

But so far we can't see the fragments. The jelly column looks uniform all the way down. There is nothing to show that DNA fragments of different size are lurking in discrete bands along its length, and nothing to show which bands contain which variety of tandem repeat. How do we make them visible? This is where the radioactive probes come in.

To make them visible you can use another cunning technique, the Southern blot, named after its inventor, Edward Southern. (Slightly confusingly, there are other techniques called the Northern blot and the Western blot, but no Mr Northern or Mr Western.) The jelly column is removed from the tube and laid out on blotting paper. The liquid in the jelly, including the DNA fragments, seeps out of the jelly into the blotting paper. The blotting paper has previously been laced with quantities of the radioactive probe for the particular tandem repeat that we are interested in. The probe molecules line up along the blotting paper, pairing precisely, by the ordinary rules of DNA, with their opposite numbers in the tandem repeats. Surplus probe molecules are washed away. Now the only radioactive probe molecules left in the blotting paper are those bound to their exact opposite numbers that seeped out of the jelly. The blotting paper is now placed on a piece of X-ray film, which is then marked by the radioactivity. So, what you see when you develop the film is a set of dark bands – another barcode. The final barcode pattern that we read on the Southern blot is a fingerprint for a person, in very much the same way as the Fraunhofer lines are a fingerprint for a star, or the formant lines are the fingerprint for a vowel sound. Indeed, the barcode from the blood looks very like Fraunhofer lines or formant lines.

The details of DNA fingerprinting techniques get quite complicated and I won't go much further. For instance, one strategy is to hit the DNA with lots of probes all at the same time. What you get then is a mixed bag of barcode stripes simultaneously. In extreme cases, the stripes merge into each other and all you get is one big smear with all possible sizes of DNA fragment represented somewhere in the genome. This is no good for identification purposes. At the other extreme, people use only one probe at a time looking at one genetic 'locus'. This 'single-locus fingerprinting' gives you nice clean bars like Fraunhofer lines. But only one or two bars per person. Even so, the chances of confusing people are small. This is because the characteristics we are talking about are not like 'brown eyes versus blue eyes', in which case lots of people would be the same. The characteristics we are measuring, remember, are lengths of tandem repeat fragments. The number of possible lengths is very large, so even single-locus fingerprinting is pretty good for identification purposes. Not quite good enough, however, so in practice forensic DNA fingerprinters usually use half a dozen separate probes. Now the chances of error are very low indeed. But we still need to talk about exactly *how* low, because people's lives or liberties might depend upon it.

First, we must return to our distinction between false positives and false negatives. DNA evidence can be used to clear an innocent suspect, or it can be made to point the finger at a guilty one. Suppose semen is recovered from the vagina of a rape victim. Circumstantial evidence leads the police to arrest a man, suspect A. Suspect A gives a blood sample and it is compared to the semen sample, using a single DNA probe to look at one tandem repeat locus. If the two are different, suspect A is in the clear. We don't even need to look at a second locus.

But what if suspect A's blood matches the semen sample at this locus? Suppose they both share the same barcode pattern, which we shall call pattern P. This is compatible with the suspect's being guilty, but it doesn't prove it. He could just happen to share pattern P with the real rapist. We must now look at some more loci. If the samples still match, what are the odds against such a match being coincidental – a false positive misidentification? This is where we have to start thinking statistically about the population at large. In theory, by taking blood from a sample of men in the population at large, we should be able to calculate the likelihood that any two men will be identical at each locus concerned. But from which section of the population do we draw our sample?

Remember our lone bearded man in the old-fashioned line-up identity parade? Here's the molecular equivalent. Suppose that, in the world at large, only one in a million men has pattern P. Does this mean that there is a million to one chance against a wrongful conviction of suspect A? No. Suspect A may belong to a minority group of people whose ancestors immigrated from a particular part of the world. Local populations often share genetic peculiarities, for the simple reason that they are descended from the same ancestors. Of the 2.5 million South African Dutch, or Afrikaners, most are descended from one shipload of immigrants who arrived from the Netherlands in 1652. As an indicator of the narrowness of this genetic bottleneck, about a million still bear the surnames of 20 of these original settlers. The Afrikaners have a much higher frequency of certain genetic diseases than the population of the world in general. According to one estimate, about 8,000 (one in 300) have the blood condition porphyria variegata, which is much rarer in the rest of the world. This is apparently because they are descended from one particular couple on the ship, Gerrit Jansz and Ariaantje Jacobs, although it is not known which one was the carrier of the (dominant) gene for the condition. (She was one of eight Rotterdam orphanage girls put on the ship to provide wives for the settlers.) In fact, the condition wasn't noticed at all before modern medicine, because its most marked symptom is a lethal reaction to certain modern anaesthetics (South African hospitals now routinely test for the gene before administering anaesthetic). Other populations often have locally high frequencies of other particular genes, for the same kind of reason. If, to return to our hypothetical court case, suspect A and the real criminal both belong to the same minority group, the likelihood of chance confusion could be dramatically greater than you'd think if you based your estimates on the population at large. The point is that the frequency of pattern P in humans at large is no longer relevant. We need to know the frequency of pattern P in the group to which the suspect belongs.

This need is nothing new. We've already seen the equivalent danger in an ordinary line-up identity parade. If the prime suspect is Chinese, it doesn't do to stand him in a line-up largely consisting of westerners. And the same kind of statistical reasoning about the background population is needed in identifying stolen goods, as well as individual suspects. I have already mentioned my jury service in the Oxford Court. In one of the three cases I sat on, a man was accused of stealing three coins from a rival numismatist. The accused had been caught with three coins in his possession which matched those lost. Counsel for the prosecution was eloquent.

Ladies and gentlemen of the jury, are we really supposed to believe that three coins, of exactly the same type as the three missing coins, would just happen to be present in the house of a rival collector? I put it to you that such a coincidence is too much to stomach.

Jurymen are not permitted to cross-examine. That was the duty of counsel for the defence, and he, though doubtless learned in the law and also eloquent, had no more clue about probability theory than the prosecutor. I wish he'd said something like this:

M'Lud, we don't know *whether the coincidence is too much to stomach, because m'learned friend has not presented us with any evidence at all as to the rarity or commonness of these three coins in the population at large. If these coins are so rare that only one in a hundred collectors in the country has any one of them, the prosecution has a good case, since the defendant was caught with three of them. If, on the other hand, these coins are as common as dirt, there is not enough evidence to convict. (To push to the extreme, three coins that I have in my pocket today, all current legal tender, are very probably the same as three coins in Your Lordship's pocket.)*

My point is that it simply never occurred to any of the legally trained minds in the court that it was relevant even to *ask* how rare these three coins were in the population at large. Lawyers can certainly add up (I once received a lawyer's bill, the last item of which was 'Time spent making out this bill') but probability theory is another matter.

I expect the coins were actually rare. If they hadn't been, the theft would not have been such a serious matter, and the prosecution presumably would never have been brought. But the jury should have been told explicitly. I remember that the question came up in the jury room, and we wished that we were allowed to go back into the court to seek clarification. The equivalent question is equally relevant in the case of DNA evidence, and it is most certainly being asked. Fortunately, provided a sufficient number of separate genetic loci are examined, the chances of misidentification – even among members of minority groups, even among family members (except identical twins) – can be reduced to genuinely very small levels, far smaller than can be achieved by any other method of identification, including eye-witness evidence.

Exactly how small the residual possibility of error is may still be open to dispute. And this is where we come to the third category of objection to DNA evidence, the just plain silly. Lawyers are accustomed to pouncing when expert witnesses seem to disagree. If two geneticists are summoned to the stand and are asked to estimate the probability of a misidentification with DNA evidence, the first may say a 1,000,000 to one while the second may say only a 100,000 to one. Pounce. 'Aha! AHA! The experts disagree! Ladies and gentlemen of the jury, what confidence can we place in a scientific method if the experts themselves can't get within a factor of ten of one another? Obviously the only thing to do is throw the entire evidence out, lock, stock and barrel.'

But, in these cases, although geneticists may be inclined to give different weightings to imponderables such as the racial subgroup effect, any disagreement between them is only over whether the odds against a wrongful identification are hyper-mega-astronomical or just plain astronomical. The odds cannot normally be lower than thousands to one, and they may well be up in the billions. Even on the most conservative estimate, the odds against wrongful identification are hugely greater than they are in an ordinary identity parade. 'M'lud, an identity parade of only 20 men is grossly unfair on my client. I demand a line-up of at least a million men!'

Expert statisticians called to give evidence on the likelihood that a conventional 20-man identity parade could yield a false identification would also disagree among themselves. Some would give the simple answer, one in 20. Under cross-examination they would then agree that it could be one in less than 20, depending upon the nature of the variation in the line-up in relation to the features of the suspect (this was the point about the lone bearded man in the line-up). But the one thing all the statisticians would agree upon is that the odds of misidentification by sheer chance are *at least* one in 20. Yet lawyers and judges are normally happy to go along with ordinary identity parades in which the suspect stands in a line of only 20 men.

After reporting the throwing out of DNA evidence in a case at London's central criminal court the Old Bailey, the *Independent* newspaper of 12 December 1992 predicted a consequent flood of appeals. The idea is that everybody at present languishing in jail, as a result of DNA identification evidence, will now be able to appeal, citing the precedent. But the flood may be even greater than the *Independent* imagines because, if this throwing out of DNA evidence is really a serious precedent for anything, it will cast doubt on *all* cases in which the odds against a chance mistake are less than thousands to one. If a witness says she 'saw' somebody and identified him in a line-up, lawyers and juries are satisfied. But the odds of mistaken identity when the human eye is involved are far greater than when the identification is done by DNA fingerprinting. If we take the precedent seriously, it ought to mean that every convicted criminal in the country will have excellent cause to appeal on grounds of mistaken identity. Even where a suspect was seen by dozens of witnesses with a smoking gun in his hand, the odds of injustice must be greater than one in 1,000,000.

A recent highly publicized case in America, where the jury were systematically confused about DNA evidence, has also become notorious for another piece of bungled probability theory. The defendant, who was known to have beaten his wife, was on trial for finally murdering her. One of the high-profile defence team, a Harvard professor of law, advanced the following argument: Statistics show that of men who beat their wives, only one in 1,000 go on to kill them. The inference that any jury might be expected to draw (indeed, were *intended* to draw) is that the defendant's beating of his wife should be discounted in the murder trial. Doesn't the evidence show overwhelmingly that a wife-beater is unlikely to turn into a wife murderer? Wrong. Doctor I. J. Hood, a professor of statistics, wrote to the scientific journal *Nature* in June 1995 to explode the fallacy. The defence lawyer's argument overlooks the additional fact that wife-killing is rare compared with wife-beating. Good calculated that if you take that minority of wives who are both beaten by their husbands and murdered by *somebody*, it is very likely indeed that the murderer will be the husband. This is the relevant way to calculate the odds because, in the case under discussion, the unfortunate wife *had* been murdered by somebody, after being beaten by her husband.

No doubt there are lawyers, judges and coroners who could benefit from a better understanding of the theory of probability. On some occasions, however, one cannot help suspecting that they understand very well and are feigning incompetence. I do not know if this was so in the case just quoted. The same suspicion is raised by Doctor Theodore Dalrymple, the (London) *Spectator*'s acerbic medical raconteur, in this typically sardonic account, from 7 January 1995, of his being called as an expert witness in a coroner's court:

. . . a wealthy and successful man I knew swallowed 200 tablets and a bottle of rum. The coroner asked me whether I thought he might have taken them by accident. I was about to answer with a ringing and confident no, when the coroner made himself a little clearer: was there even a one in a million chance he had taken them by accident?

'Er, well, I suppose so,' I replied. The coroner (and the man's family) relaxed, an open verdict was returned, the family was £750,000 the richer and an insurance company the poorer by an equivalent sum, at least until it put my premium up.

The power of DNA fingerprinting is an aspect of the general power of science that makes some people fear it. It is important not to exacerbate such fears by claiming too much or trying to move too fast. Let me end this rather technical chapter by returning to society and an important and difficult decision that we must collectively make. I would normally fight shy of discussing a topical issue for fear of going out of date, or a local one for fear of being parochial, but the question of a national DNA database is starting to preoccupy most nations in their different ways, and it is bound to become more pressing in the future.

It would in theory be possible to keep a national database of DNA sequences from every man, woman and child in the country. Then, whenever a sample of blood, semen, saliva, skin or hair was found at the scene of a crime, the police would not have to locate a suspect by other means before comparing his DNA with the sample. They could simply do a computer search of the national database. The very suggestion elicits howls of protest. It would be an infringement of individual liberty. It's the thin end of the wedge. A giant step towards a police state. I have always been a little puzzled about why people *automatically* react so strongly against suggestions such as these. If I examine the matter dispassionately, I think that, on balance, I come out against it. But it is not something to condemn out of hand without even looking at the pros and cons. So let us do so.

If the information is guaranteed to be used only for catching criminals, it is hard to see why anybody who is not a criminal should object. I am aware that plenty of activists for civil liberties will still object in principle. But I genuinely don't understand why, unless we want to protect the rights of criminals to perform crimes without detection. I also see no good reason against a national database of conventional, inkpad fingerprints (except the practical one that, unlike with DNA, it is hard to do an automatic computer search of conventional fingerprints). Crime is a serious problem which diminishes the quality of life for everybody except the criminals (perhaps even them: presumably there is nothing to stop a burglar's house being burgled). If a national DNA database would significantly help the police to catch criminals, the objections had better be good ones to outweigh the benefits.

Here's an important caution, though, to begin with. It's one thing to use DNA evidence, or mass-screening identification evidence of any kind, to corroborate a suspicion that the police have already reached on other grounds. It's quite another matter to use it to arrest anybody in the country who matches the sample. If there is a certain low probability of *coincidental* resemblance between, say, a semen sample and the blood of an innocent individual, the probability that that individual will *also* be falsely suspected on independent grounds is obviously far lower. So the technique of simply searching the database and arresting the one person who matches the sample is significantly more likely to lead to injustice than a system which requires other grounds for suspicion first. If a sample from the scene of a crime in Edinburgh happens to match my DNA, should the police be allowed to hammer on my door in Oxford and arrest me on no other evidence? I think not, but it is worth remarking that the police already do something equivalent with facial features, when they release to the national newspapers an Identikit picture, or a snapshot taken by a witness, and invite people from all over the country to telephone them if they 'recognize' the face. Once again, we must beware of our natural tendency to trust facial recognition above all other kinds of individual identification.

Setting crime aside, there is a real danger of the information in the national DNA database falling into the wrong hands. I mean into the hands of those who wish to use it not for catching criminals but for other purposes, perhaps connected with medical insurance or blackmail. There are respectable reasons why people with no criminal intent at all might not wish their DNA profile to be known, and it seems to me that their privacy should be respected. For instance, a significant number of individuals who believe they are the father of a particular child are not. Equally, a significant number of children believe somebody to be their real father who is not. Anyone with access to the national DNA database might discover the truth, and the result could be huge emotional distress, marital breakdown, nervous breakdown, blackmail, or worse. There may be some who feel that the truth should always out, however painful, but I think a good case could be made that the sum total of human happiness would not be enhanced by a sudden outburst of revelations about everybody's true paternity.

Then there are the medical and insurance issues. The whole life insurance business depends upon the inability to forecast exactly when somebody will die. As Sir Arthur Eddington said: 'Human life is proverbially uncertain; few things are more certain than the solvency of a life-insurance company.' We all pay our premiums. Those of us who die later than expected subsidize (the heirs of) those who die earlier than expected. Insurance companies already make statistical guesses which partially subvert the system by enabling them to charge high-risk clients larger premiums. They send a doctor to listen to our hearts, take our blood pressure and investigate our smoking and drinking habits. If actuaries knew exactly when we were all going to die, life insurance would become impossible. In principle, a national DNA database, if actuaries could get their hands on it, might lead us closer to this unfortunate outcome. An extreme could be reached where the only kind of death risk that could be insured against would be pure accident.

Similarly, people screening job applicants, or applicants for places at university, could use DNA information in ways that many of us might find undesirable. Some employers already use dubious methods such as graphology (analysis of handwriting as a supposed guide to character or aptitude). Unlike the case of graphology, there is good reason to think that DNA information might be genuinely useful for judging abilities. But still, I would be one of many who would be disturbed if selection panels made use of DNA information, at least if they did so secretly.

One of the general arguments against national databases of any kind is the 'What if it fell into the hands of a Hitler?' argument. On the face of it, it is not clear how an evil government would benefit from a database of true information about people. They are so adept at using false information, one might say, why should they bother to abuse true information? In the case of Hitler, however, there is the point about his campaign against Jews and others. Although it is not true that you can recognize a Jew from his DNA, there are particular genes which are characteristic of people whose ancestors come from certain regions of, say, central Europe, and there are statistical correlations between possession of certain genes and being Jewish. It seems undeniable that, if Hitler's regime had had a national DNA database at their disposal, they would have found terrible ways to abuse it.

Are there ways to safeguard society from these potential ills, while retaining the benefit of helping to catch criminals? I'm not sure. I think it might be difficult. You could protect honest citizens against insurance companies and employers by restricting the national database to non-coding regions of the genome. The database would refer only to tandem repeat areas of the genome, not genes that actually do anything. This would prevent actuaries working out our life expectancy and talent scouts second-guessing our abilities. But it would do nothing to protect us against discovering (or against blackmailers discovering) truths about paternity that we might prefer not to know. Quite the contrary. The identification of Josef Mengele's bones from his son's blood was entirely based upon tandem repeat DNA. I see no easy answer to this objection except to say that, as DNA testing becomes easier, it will increasingly be possible to discover paternity in any case, without recourse to a national database. A man who suspects that 'his' child is not really his could already take the child's blood and have it compared with his own. He wouldn't need a national database.

Not just in courts of law, the decisions of commissions of inquiry and other bodies charged with discovering what happened in some incident or accident frequently turn upon scientific matters. Scientists are called as expert witnesses on factual matters: on the technicalities of metal fatigue, on the infectivity of mad cow disease, and so on. Then, having delivered their expertise, the scientists are dismissed so those charged with the serious business of actually making the decisions can get on with it. The implication is that scientists are good at discovering detailed facts but others, often lawyers or judges, are better qualified to integrate them and recommend what needs to be done. On the contrary, a good case can be made that scientific ways of thinking are valuable, not just for assembling the detailed facts but for reaching the final verdict. When there has been an air crash, say, or a disastrous football riot, a scientist might be better qualified to chair the inquiry than a judge, not because of what scientists know, but because of the methods they use to find things out and make decisions.

The case of DNA fingerprinting suggests that lawyers would be better lawyers, judges better judges, parliamentarians better parliamentarians and citizens better citizens if they knew more science and, more to the point, if they reasoned more like scientists. This is not only because scientists value reaching the truth above winning a case. Judges, and decision-takers in general, might be better decision-takers if they were more adept in the arts of statistical reasoning and probability assessment. This point will resurface in the next two chapters, which deal with superstition and the so-called paranormal.

6

HOODWINK'D WITH FAERY FANCY

Credulity is the man's weakness, but the child's strength.

CHARLES LAMB, *Essays of Elia* (1823)

We have an appetite for wonder, a poetic appetite, which real science ought to be feeding but which is being hijacked, often for monetary gain, by purveyors of superstition, the paranormal and astrology. Resonant phrases like 'the Fourth House of the Age of Aquarius', or 'Neptune went retrograde and moved into Sagittarius' whip up a bogus romance which, to the naïve and impressionable, is almost indistinguishable from authentic scientific poetry: 'The Universe is lavish beyond imagining' for example, from Carl Sagan and Ann Druyan's *Shadows of Forgotten Ancestors* (1992); or, out of the same book (after describing how the solar system condensed out of a spinning disc), 'The disk is rippling with possible futures.' In another book, Carl Sagan remarked,

How is it that hardly any major religion has looked at science and concluded, 'This is better than we thought! The Universe is much bigger than our prophets said, grander, more subtle, more elegant'? Instead they say, 'No, no, no! My god is a little god, and I want him to stay that way.' A religion, old or new, that stressed the magnificence of the Universe as revealed by modern science might be able to draw forth reserves of reverence and awe hardly tapped by the conventional faiths.

Pale Blue Dot (1995)

In so far as traditional religions are in decline in the West, their place seems to be taken not by science, with its clearer-sighted, grander vision of the cosmos, so much as by the paranormal and astrology. One might have hoped that by the end of this most scientifically successful of all centuries science would have been incorporated into our culture and our aesthetic sense risen to meet its poetry. Without reviving the mid-century pessimism of C. P. Snow, I reluctantly find that, with only two years to run, these hopes are not realized. Astrology books outsell astronomy books. Television beats a path to the doors of second-rate conjurors masquerading as psychics and clairvoyants. This chapter examines superstition and gullibility, trying to explain them and the ease with which they can be exploited. Chapter 7 then advocates simple statistical thinking as an antidote to the paranormal disease. We begin with astrology.

On 27 December 1997, one of Britain's largest circulation national newspapers, the *Daily Mail*, devoted its main front-page story to astrology under the banner headline '1998: The Dawn of Aquarius'. One feels almost grateful when the article goes on to concede that the Hale Bopp comet was not the *direct* cause of Princess Diana's death. The paper's highly paid astrologer tells us that 'slow-moving, powerful Neptune' is about to join 'forces' with the equally powerful Uranus as it moves into Aquarius. This will have dramatic consequences:

... the Sun is rising. And the comet has come to remind us that this Sun is not a physical sun but a spiritual, psychic, inner sun. It does not, therefore, have to obey the law of gravity. It can come over the horizon more swiftly if enough people rise to greet and encourage it. And it can dispel the darkness the moment it appears.

How can people find this meaningless pap appealing, especially in the face of the real universe as revealed by astronomy?

On a moonless night when 'the stars look very cold about the sky', and the only clouds to be seen are the glowing smudges of the Milky Way, go out to a place far from street light pollution, lie on the grass and gaze up at the sky. Superficially you notice constellations, but a constellation's pattern means no more than a patch of damp on the bathroom ceiling. Note, accordingly, how little it means to say something like 'Neptune moves into Aquarius'. Aquarius is a miscellaneous set of stars all at different distances from us which are unconnected with each other except that they constitute a (meaningless) pattern when seen from a certain (not particularly special) place in the galaxy (here). A constellation is not an entity at all, and so not the kind of thing that Neptune, or anything else, can sensibly be said to 'move into'.

The shape of a constellation, moreover, is ephemeral. A million years ago our *Homo erectus* ancestors gazed out nightly (no light pollution then, unless it came from that species' brilliant innovation, the camp fire) at a set of very different constellations. A million years hence, our descendants will see yet other shapes in the sky and we already know exactly how these will look. This is the sort of detailed prediction that astronomers, but not astrologers, can make. And – again by contrast with astrological predictions – it will be correct.

Because of light's finite speed, when you look at the great galaxy in Andromeda you are seeing it as it was 2.3 million years ago and *Australopithecus* stalked the high veldt. You are looking back in time. Shift your eyes a few degrees to the nearest bright star in the constellation of Andromeda and you see Mirach, but much more recently, as it was when Wall Street crashed. The sun, when you witness its colour and shape, is only eight minutes ago. But point a large telescope at the Sombrero galaxy and you behold a trillion suns as they were when your tailed ancestors peered shyly through the canopy and India collided with Asia to raise the Himalayas. A collision on a larger scale, between two galaxies in Stephan's Quintet, is shown to us at a time when on earth dinosaurs were dawning and the trilobites fresh dead.

Name any event in history and you will find a star out there whose light gives you a glimpse of something happening during the year of that event. Provided you are not a very young child, somewhere up in the night sky you can find your personal birth star. Its light is a thermonuclear glow that heralds the year of your birth. Indeed, you can find quite a few such stars (about 40 if you are 40; about 70 if you are 50; about 175 if you are 80 years old). When you look at one of your birth year stars, your telescope is a time machine letting you witness thermonuclear events that are actually taking place during the year you were born. A pleasing conceit, but that is all. Your birth star will not deign to tell anything about your personality, your future or your sexual compatibilities. The stars have larger agendas in which the preoccupations of human pettiness do not figure.

Your birth star, of course, is yours for only this year. Next year you must look to the surface of a larger sphere one light year more distant. Think of this expanding sphere as a radius of good news, the news of your birth broadcast steadily outwards. In the Einsteinian universe in which most physicists now think we live, nothing can in principle travel faster than light. So, if you are 50 years old, you have a personal news bubble of 50 light years' radius. Within that sphere (of a little more than a thousand stars) it is *in principle* possible (although obviously not in practice) for news of your existence to have permeated. Outside that sphere you might as well not exist; in an Einsteinian sense you do not exist. Older people have larger existence spheres than younger people, but nobody's existence extends to more than a tiny fraction of the universe. The birth of Jesus may seem an ancient and momentous event to us as we reach his second millenary. But the news is so recent on this scale that, even in the most ideal circumstances, it could in principle have been proclaimed to less than one 200 million millionth of the stars in the universe. Many, if not most, of the stars out there will be orbited by planets. The numbers are so vast that probably some of them have life forms, some have evolved intelligence and technology.

Yet the distances and times that separate us are so great that thousands of life forms could independently evolve and go extinct without it being possible for any to know of the existence of any other.

In order to make my calculations about numbers of birth stars, I assumed that the stars are spaced, on average, about 7.6 light years apart. This is approximately true of our local region of the Milky Way galaxy. It seems an astonishingly low density (about 440 cubic light years per star), but it is actually high by comparison with the density of stars in the universe as a whole, where space lies empty between the galaxies. Isaac Asimov has a dramatic illustration: it is as if all the matter of the universe were a single grain of sand, set in the middle of an empty room 20 miles long, 20 miles wide and 20 miles high. Yet, at the same time, it is as if that single grain of sand were pulverized into a thousand million million million fragments, for that is approximately the number of stars in the universe. These are some of the sobering facts of astronomy, and you can see that they are beautiful.

Astrology, by comparison, is an aesthetic affront. Its pre-Copernican dabblings demean and cheapen astronomy, like using Beethoven for commercial jingles. It is also an insult to the science of psychology and the richness of human personality. I am talking about the facile and potentially damaging way in which astrologers divide humans into 12 categories. Scorpios are cheerful, outgoing types while Leos, with their methodical personalities, go well with Libras (or whatever it is). My wife Lalla Ward recalls an occasion when an American starlet approached the director of the film they were both working on with a 'Gee, Mr Preminger, what sign are you?' and received the immortal rebuff, in a thick Austrian accent, 'I am a Do Not Disturrrb sign.'

Personality is a real phenomenon and psychologists have had some success in developing mathematical models to handle its variation in many dimensions. The initially large number of dimensions can be mathematically collapsed into fewer dimensions with measurable, and for some purposes conscionable, loss in predictive power. These fewer derived dimensions sometimes correspond to the dimensions that we intuitively think we recognize – aggressiveness, obstinacy, affectionateness and so on. Summarizing an individual's personality as a point in multidimensional space is a serviceable approximation whose limitations can be stated. It is a far cry from any mutually exclusive categorization, and certainly far from the preposterous fiction of newspaper astrology's 12 dumpbins. It is based upon genuinely relevant data about people themselves, not their birthdays. The psychologist's multidimensional scaling can be useful in deciding whether a person is suited to a particular career, or a proposed couple to each other. The astrologer's 12 pigeonholes are, if nothing worse, a costly and irrelevant distraction.

Moreover, they sit oddly with our current strong taboos, and laws, against discrimination. Newspaper readers are schooled to regard themselves and their friends and colleagues as Scorpios or Libras or one of the other 12 mythic 'signs'. If you think about it for a moment, isn't this a form of discriminatory labelling rather like the cultural stereotypes which many of us nowadays find objectionable? I can imagine a Monty Python sketch in which a newspaper publishes a daily column something like this:

Germans: It is in your nature to be hard-working and methodical, which should serve you well at work today. In your personal relationships, especially this evening, you will need to curb your natural tendency to obey orders.

Spaniards: Your Latin hot blood may get the better of you, so beware of doing something you might regret. And lay off the garlic at lunch if you have romantic aspirations in the evening.

Chinese: Inscrutability has many advantages, but it may be your undoing today . . .

British: Your stiff upper lip may serve you well in business dealings, but try to relax and let yourself go in your social life.

And so on through 12 national stereotypes. No doubt the astrology columns are less offensive than this, but we should ask ourselves exactly where the difference lies. Both are guilty of facile discrimination, dividing humanity up into exclusive groups based upon no evidence. Even if there were evidence of some slight statistical effects, both kinds of discrimination encourage prejudiced handling of people as *types* rather than as individuals. You can already see advertisements in lonely hearts columns that include phrases like 'No Scorpios' or 'Tauruses need not apply'. Of course this is not as bad as the infamous 'No blacks' or 'No Irish' notices, because astrological prejudice doesn't consistently pick on some star signs more than others, but the principle of discriminatory stereotyping – as opposed to accepting people as individuals – remains.

There could even be sad human consequences. The whole point of advertising in lonely hearts columns is to increase the catchment area for meeting sexual partners (and indeed the circle provided by the workplace and by friends of friends is often meagre and needs enriching). Lonely people, whose life might be transformed by a longed-for compatible friendship, are encouraged to throw away wantonly and pointlessly up to eleven twelfths of the available population. There are some vulnerable people out there and they should be pitied, not deliberately misled.

On an apocryphal occasion a few years ago, a newspaper hack who had drawn the short straw and been told to make up the day's astrological advice relieved his boredom by writing under one star sign the following portentous lines: 'All the sorrows of yesteryear are as nothing compared to what will befall you today.' He was fired after the switchboard was jammed with panic-stricken readers, pathetic testimony to the simple trust people can place in astrology.

In addition to anti-discrimination legislation, we have laws designed to protect us from manufacturers making false claims for their products. The law is not invoked in defence of simple truth about the natural world. If it were, astrologers would provide as good a test case as could be desired. They make claims to forecast the future and divine personal foibles, and they take payment for this, as well as for professional advice to individuals on important decisions. A pharmaceuticals manufacturer who marketed a birth control pill that had not the slightest demonstrable effect upon fertility would be prosecuted under the Trade Descriptions Act, and sued by customers who found themselves pregnant. Once again it feels like over-reaction, but I cannot actually work out why professional astrologers are not arrested for fraud as well as for incitement to discrimination.

The London *Daily Telegraph* of 18 November 1997 reported that a self-styled exorcist who had persuaded a gullible teenage girl to have sex with him on the pretext of driving evil spirits from her body had been jailed for 18 months the day before. The man had shown the young woman some books on palmistry and magic, then told her that she was 'jinxed: someone had put bad luck on her'. In order to exorcise her, he explained, he needed to anoint her all over with special oils. She agreed to take all her clothes off for this purpose. Finally, she copulated with the man when he told her that this was necessary 'to get rid of the spirits'. Now, it seems to me that society cannot have it both ways. If it was right to jail this man for exploiting a gullible young woman (she was above the legal age of consent), why do we not similarly prosecute astrologers who take money off equally gullible people; or 'psychic' diviners who con oil companies into parting with shareholders' money for expensive 'consultations' on where to drill? Conversely, if it be protested that fools should be free to hand over their money to charlatans if they choose, why shouldn't the sexual 'exorcist' claim a similar defence, invoking the young woman's freedom to give her body for the sake of a ritual ceremony in which, at the time, she genuinely believed?

There is no known physical mechanism whereby the position of distant heavenly bodies at the moment of your birth could exert any causal influence on your nature or your destiny. This does not rule out the possibility of some unknown physical influence. But we need bother to think about such a physical influence only if somebody can produce any evidence that the movements of planets against the backdrop of constellations actually has the slightest influence on human affairs. No such evidence has ever stood up to proper investigation. The vast majority of scientific studies of astrology have yielded no positive results whatever. A (very) few studies have suggested (weakly) a statistical correlation between star 'sign' and character. These few positive results turned out to have an interesting explanation. Many people are so well versed in star sign lore that they know which characteristics are expected of them. They then have a small tendency to live up to these expectations – not much, but enough to produce the very slight statistical effects observed.

A minimal test that any reputable method of diagnosis or divining ought to pass is that of *reliability*. This is not a test of whether it actually works, merely a test of whether different practitioners confronted with the same evidence (or the same practitioner confronted with the same evidence twice) agree. Although I don't think astrology works, I really would have expected high reliability scores in this sense of self-consistency. Different astrologers, after all, presumably have access to the same books. Even if their verdicts are wrong, you'd think their methods would be systematic enough at least to agree in producing the *same* wrong verdicts! Alas, as shown in a study by G. Dean and colleagues, they don't even achieve this minimal and easy benchmark. For comparison, when different assessors judged people on their performance in structured interviews, the correlation coefficient was greater than 0.8 (a correlation coefficient of 1.0 would represent perfect agreement, –1.0 would represent perfect disagreement, 0.0 would represent complete randomness or lack of association; 0.8 is pretty good). Against this, in the same study, the reliability coefficient for astrology was a pitiable 0.1, comparable to the figure for palmistry (0.11), and indicating near total randomness. However wrong astrologers may be, you'd think that they would have got their act together to the extent of at least being *consistent.* Apparently not. Graphology (handwriting analysis) and Rorschach (inkblot) analyses aren't much better.

The job of astrologer requires so little training or skill that it is often handed out to any junior reporter with time on his hands. The journalist Jan Moir relates in the *Guardian* on 6 October 1994 that, 'My very first job in journalism was writing horoscopes for a stable of women's magazines. It was the office task always given to the newest recruit because it was so stupid and so easy that even a wet-eared geek like me could do it.' Similarly, when he was a young man the conjuror and rationalist James Randi took a job, under the pseudonym Zo-ran, as astrologer on a Montreal newspaper. Randi's method of working was to take old astrology magazines, cut out their forecasts with scissors, stir them around in a hat, paste them at random under the 12 'signs', then publish them as his own 'forecasts'. He describes how he overheard a pair of office workers in their lunch break in a café eagerly scanning 'Zo-ran's' column in the paper.

They squealed with delight on seeing their future so well laid out, and in response to my query said that Zo-ran had been 'right smack on' last week. I did not identify myself as Zo-ran . . . Reaction in the mail to the column had been quite interesting, too, and sufficient for me to decide that many people will accept and rationalize almost any pronouncement made by someone they believe to be an authority with mystic powers. At this point, Zo-ran hung up his scissors, put away the paste pot, and went out of business. *Flim-Flam* (1992)

There is evidence from questionnaire research that many people who read daily horoscopes don't really believe them. They state that they read them only as 'entertainment' (their taste in what constitutes entertaining fiction is evidently different from mine). But significant numbers of people really do believe and act upon them including, according to alarming and apparently authentic reports, Ronald Reagan during his time as president. Why is anybody impressed by horoscopes?

First, the forecasts, or character-readings, are so bland, vague and general that they fit almost anybody and any circumstance. People normally read only their own horoscope in the newspaper. If they forced themselves to read the other 11 they'd be far less impressed with the accuracy of their own. Second, people remember the hits and overlook the misses. If there is one sentence in a paragraph-long horoscope which seems to strike home, you notice that particular sentence while your eye skims unseeingly over all the other sentences. Even if people do notice a strikingly wrong forecast, it is quite likely to be chalked up as an interesting exception or anomaly rather than as an indication that the whole thing might be baloney. Thus David Bellamy, a popular television scientist (and genuine conservationist hero), confided in *Radio Times* (that once-respected organ of the BBC) that he has the 'Capricorn caution' over certain things, but mostly he puts his head down and charges like a real goat. Isn't that interesting? Well, I do declare, it just bears out what I always say: it's the exception that proves the rule! Bellamy himself presumably knew better, and was just going along with the common tendency among educated people to indulge astrology as a bit of harmless entertainment. I doubt if it is harmless, and I wonder whether people who describe it as entertaining have ever actually been entertained by it.

'Mum Gives Birth to 8 lb Kitten' is a typical headline from a paper called *Sunday Sport* which, like its American equivalents such as the *National Enquirer* (with a circulation of 4 million), is entirely devoted to printing ludicrously tall stories as if they were fact. I once met a woman who was employed full time to invent these stories for an American publication of this kind, and she told me she and her colleagues vied with each other to see who could get away with the most outrageously ridiculous items. It turned out to be an empty competition, because there doesn't seem to be any limit to what people will believe if only they see it in print.

On the page following the eight-pound kitten story, the *Sunday Sport* carried an article about a magician who couldn't stand his wife's nagging so he turned her into a rabbit. In addition to this pandering to the prejudiced cliché of the nagging wife, the same issue of the paper added a xenophobic flavour to its fantasies: 'Mad Greek turns Boy into Kebab'. Other well-loved stories from these papers include 'Marilyn Monroe Comes Back as a Lettuce' (complete with green-tinted photograph of the late screen goddess's face nestling in the heart of a fresh young vegetable) and 'Statue of Elvis Found on Mars'.

Sightings of a resurrected Elvis Presley are numerous. The cult of Elvis, with its treasured toenails and other relics, its icons and its pilgrimages, is well on the way to becoming a fully fledged new religion, but it will have to look to its laurels if it is not to be overtaken by the younger cult of Princess Diana. The crowds queuing to sign the condolence book after her death in 1997 reported to journalists that her face was clearly seen through a window, peering out of an old portrait hanging on a wall. As in the case of the Angel of Mons, who appeared to soldiers during the darkest days of the First World War, numerous eye-witnesses 'saw' the spectre of Diana, and the story spread like a bushfire among the keening crowds, whipped up as they were by the tabloid newspapers.

Television is an even more powerful medium than the newspapers, and we are in the grip of a near epidemic of paranormal propaganda on television. In one of the more notorious examples of recent years in Britain, a faith healer claimed to be the receptacle for the soul of a 2,000-year dead doctor called Paul of Judea. With not a whisper of critical inquiry, the BBC devoted an entire half-hour programme to promoting his fantasy as fact. Afterwards, I clashed with the commissioning editor of this programme, in a public debate on 'Selling Out to the Supernatural' at the 1996 Edinburgh Television Festival. The editor's main defence was that the man was doing a good job healing his patients. He seemed genuinely to feel that this was all that mattered. Who cares whether reincarnation really happens, as long as the healer can bring some comfort to his patients? For me, the real crusher came in a publicity hand-out that the BBC released to accompany the show. Among those acknowledged for advice, and listed as overseeing the content, was none other than . . . Paul of Judea. It is one thing for people to be shown on their screens the eccentric beliefs of a psychotic or fraudulent individual. Perhaps this is entertainment – comedy even, although I find it as objectionable as laughing at a fairground freak show, or the current vogue in America for setting up violent marital disputes on television. But it is quite another thing for the BBC to lend the weight of its long built-up reputation by appearing to *accept* the fantasy at face value in the billing.

A cheap but effective formula for paranormal television is to employ ordinary conjurors, but repeatedly tell the audience they are not conjurors but genuinely supernatural. In an added display of cynical contempt for the viewer's IQ, these acts are subjected to less control and precaution than a performing magician normally would be. *Bona fide* conjurors at least go through the motions of demonstrating that there is nothing up their sleeve, no wires under the table. When an artist is billed as 'paranormal' he is excused even this perfunctory handicap.

Let me describe an actual item, a telepathy act, from Carlton television's recent series, *Beyond Belief*, produced and presented by David Frost, a veteran British television personality whom some government saw fit to knight and whose imprimatur, therefore, carries weight with viewers. The performers were a father-and-son team from Israel in which the blindfolded son would see 'through his father's eyes'. A randomizing device was spun, and a number came up. The father stared fixedly at it, clenching and unclenching his fists under the strain, and asked his son in a strangled shout whether he could do it. 'Yes, I think so,' croaked the son. And, of course, he got the number right. Wild applause. How astounding! And don't forget, viewers, this is all live TV, and it is *factual programming*, not fiction like *The X-Files*.

What we have witnessed is nothing more than a familiar, rather mediocre conjuring trick, a favourite in the music halls dating back at least to Signor Pinetti in 1784. There are many simple codes by which the father could have transmitted a number to his well-rehearsed son. The word-count in his apparently innocent shout of 'Can you do it, son?' is one possibility. Instead of goggling with amazement, David Frost should have tried the simple experiment of gagging the father as well as blindfolding the son. The only difference from an ordinary conjuring show is that a reputable television company has billed it as 'paranormal'.

Most of us don't know how conjurors do their tricks. I'm often dumbfounded by them. I don't understand how they pull rabbits out of hats or saw boxes in half without harming the lady inside. But we all know that there's a perfectly good explanation which the conjuror could tell us if he wanted to but, understandably enough, he doesn't. So why should we think it a genuine miracle when exactly the same kind of trick has the 'paranormal' label slapped on it by a television company?

Then there are those performers who seem to 'sense' that somebody in the audience had a loved one whose name began with M, owned a Pekinese, and died of something to do with the chest: 'clairvoyants' and 'mediums' with apparent knowledge that they 'couldn't have got by any normal means'. I haven't space to go into details, but the trick is well known to conjurors under the name 'cold reading'. It's a subtle combination of knowing what's common (many people die of heart failure or lung cancer) and fishing for clues (people involuntarily give the game away when you are getting warm), aided by the audience's willingness to remember hits and overlook misses. Cold readers also often use narks, who eavesdrop conversations as the audience walks into the theatre, or even cross-examine people, and then report to the performer in his dressing room before the show.

If a paranormalist could really give a properly researched demonstration of telepathy (precognition, psychokinesis, reincarnation, perpetual motion, whatever it is) he would be the discoverer of a totally new principle, unknown to physical science. The discoverer of the new energy field that links mind to mind in telepathy, or of the new fundamental force that moves objects without trickery around a tabletop, deserves a Nobel Prize, and would probably get one. If you are in possession of this revolutionary secret of science, why waste it on gimmicky television entertainment? Why not prove it properly and be hailed as the new Newton? Of course, we know the real answer. You can't do it. You are a fake. But, thanks to gullible or cynical television producers, a well-heeled fake.

Having said that, some 'paranormalists' are skilled enough to fool most scientists, and the people best qualified to see through them are not scientists but other conjurors. This is why the most famous psychics and mediums regularly make excuses and refuse to go on stage if they hear that the front row of the audience is filled with professional conjurors. Various good conjurors, including James Randi in America and Ian Rowland in Britain, put on shows in which they publicly duplicate the 'miracles' of famous paranormalists – then explain to the audience that they are only tricks. The Rationalists of India are dedicated young conjurors who travel round the villages unmasking so-called 'holy men' by duplicating their 'miracles'. Unfortunately, some people *still* believe in miracles, even after the trickery has been explained. Others fall back on desperation: 'Well, maybe Randi does it by trickery,' they say, 'but that doesn't mean others aren't doing real miracles.' To this, Ian Rowland memorably retorted: 'Well, if they *are* doing miracles, they're doing it the hard way!'

There is a great deal of money to be made out of misleading the gullible. A normal workaday conjuror could not ordinarily hope to break out of the children's party market and hit nationwide television. But if he passes his tricks off as genuinely supernatural, it may be another matter. The television companies are eager collaborators in the deception. It is good for ratings. Instead of applauding politely when a competent conjuring trick has been performed, presenters gasp histrionically and lead viewers on to believe that they have witnessed something that defies the laws of physics. Disturbed people recount their fantasies of ghosts and poltergeists. But instead of sending them off to a good psychiatrist, television producers eagerly sign them up and then hire actors to perform dramatic reconstructions of their delusions – with predictable effects on the credulity of large audiences.

I am in danger of being misunderstood, and it is important that I confront this danger. It would be too easy to claim complacently that our present scientific knowledge is all that there is to know – that we can be sure astrology and spooks are rubbish, without further discussion, simply because existing science cannot explain them. Is it, after all, so obvious that astrology is a load of bunk? How do I know that a human mother didn't give birth to an eight-pound kitten? How can I be sure that Elvis Presley has not ascended in glorious resurrection, leaving an empty tomb? Stranger things have happened. Or, to be more precise, things that we accept as commonplace, such as radio, would have seemed, to our ancestors, every bit as far-fetched as spectral visitation. To us, a mobile telephone may be no more than an antisocial nuisance on trains. But to our ancestors from the nineteenth century, when trains were new, a mobile telephone would have seemed pure magic. As Arthur C. Clarke, the distinguished science fiction writer and evangelist for the limitless power of science and technology, has said, 'Any sufficiently advanced technology is indistinguishable from magic.' This has been called Clarke's Third Law, and I shall return to it.

William Thomson, first Lord Kelvin, was one of the most distinguished and influential of nineteenth-century British physicists. He was a thorn in Darwin's side because he 'proved', with massive authority but, as we now know, even more massive error, that the earth was too young for evolution to have occurred. He is also credited with the following three confident predictions: 'Radio has no future'; 'Heavier than air flying machines are impossible'; 'X-rays will prove to be a hoax.' Here was a man who took scepticism to the point where he courted – and earned – the ridicule of future generations. Arthur C. Clarke himself, in his visionary book *Profiles of the Future* (1982), tells similar cautionary tales and awful warnings of the dangers of dogmatic scepticism. When Edison announced that he was working on electric light in 1878, a British parliamentary commission was set up to investigate whether there was anything in it. The committee of experts reported that his fantastic idea (what we now know as the light bulb) was 'good enough for our transatlantic friends . . . but unworthy of the attention of practical or scientific men'.

Lest this sound like an anti-British series of stories, Clarke also quotes two distinguished American scientists on the subject of aeroplanes. The astronomer Simon Newcomb was unlucky enough to make the following remark only just before the Wright brothers' famous exploit in 1903:

The demonstration that no possible combination of known substances, known forms of machinery and known forms of force, can be united in a practical machine by which men shall fly long distances through the air, seems to the writer as complete as it is possible for the demonstration of any physical fact to be.

Another noted American astronomer, William Henry Pickering, categorically stated that, although heavier than air flying machines were *possible* (he had to say that because the Wright brothers had by then already flown) they could never be a serious practical proposition:

The popular mind often pictures gigantic flying machines speeding across the Atlantic and carrying innumerable passengers in a way analogous to our modern steamships . . . It seems safe to say that such ideas must be wholly visionary, and even if a machine could get across with one or two passengers the expense would be prohibitive . . . Another popular fallacy is to expect enormous speed to be obtained.

Pickering goes on to 'prove' by means of authoritative calculations on the effects of air resistance that an aeroplane could never travel faster than the express trains of his day. On the face of it, the 1943 remark of Thomas J. Watson, head of IBM, 'I think there is a world market for maybe five computers' sounds similar. But this is unfair. Watson was surely forecasting that computers would become ever larger, and in this he was wrong; however, he was not downgrading the importance of the computer in the future, the way Kelvin and the others were downgrading air travel.

Those banana skin stories are, indeed, awful warnings of the dangers of an over-zealous scepticism. Dogmatic disbelief of anything that seems unfamiliar or unexplained is not a virtue. What, then, is the difference between this and my avowed scepticism of astrology, reincarnation and the resurrection of Elvis Presley? How are we to know when scepticism is justified, and when it is dogmatic, intolerant short-sightedness?

Let's think about a spectrum of stories that people might tell us and meditate on how sceptical we ought to be of them. At the lowest level are stories that might be true, and might not be true, but that we have no particular reason to doubt. In Evelyn Waugh's *Men at Arms* (1952), the comic character Apthorpe frequently speaks to the narrator, Guy Crouchback, of his two aunts, one who lives in Peterborough, the other in Tunbridge Wells. On his deathbed, Apthorpe finally confesses that in fact he has only one aunt. Which one did he invent, Guy Crouchback asks. 'The one at Peterborough, of course.' 'You certainly took me in thoroughly.' 'Yes, it was a good joke, wasn't it?'

No, Apthorpe's was not a good joke, and it is precisely this that makes Evelyn Waugh's joke at Apthorpe's expense funny. There are, no doubt, many elderly ladies residing in Peterborough, and if a man tells you he has an aunt there you have no particular reason to disbelieve him. Unless he has some specific motive for lying to you, you might as well believe him, though if a great deal hangs on it you'd be wise to check the evidence. But now suppose somebody tells you that his aunt can levitate herself by meditation and will-power. She sits cross-legged, you are told, and by thinking beautiful thoughts and intoning a mantra she raises herself above the ground and stays there, hovering. Why be any more sceptical than you would be if a man simply told you that his aunt exists in Peterborough, for in both cases you have the word of a claimed eye-witness?

The obvious reply is that levitation by will-power is not explicable by science. But that just means present day science. It brings us straight back to Clarke's Third Law, and the important point that any era's science doesn't have all the answers and will be superseded. Maybe, some day in the future, physicists will fully understand gravity and build an anti-gravity machine. It is conceivable that levitating aunts will become as commonplace to our descendants as jet planes are to us. Does Clarke's Third Law then entitle us to believe any and every yarn that folk may spin about apparent miracles? If a man claims to have witnessed his aunt in cross-legged levitation, or a Turk zooming over the minarets on a magic carpet, should we swallow his story on the grounds that those of our ancestors who doubted the possibility of radio turned out to be wrong? No, *of course* these are not sufficient grounds for believing in levitation or magic carpets. But why not?

Clarke's Third Law does not work in reverse. Given that 'Any sufficiently advanced technology is indistinguishable from magic', it does *not* follow that 'Any magical claim that anybody may make at any time is indistinguishable from a technological advance that will come in the future.' Yes, there have been occasions when authoritative sceptics have come away with egg on their pontificating faces. But a far greater number of magical claims have been made and never vindicated. A few things that would surprise us today will come true in the future. But far more things that would surprise us today will *not* come true in the future. The trick is to sort out the minority from the rubbish – from claims that will forever remain in the realm of fiction and magic.

If faced with an amazing or miraculous story, we can begin by asking ourselves whether our informant has a motive to lie. Or we can assess his credentials in other ways. I recall an entertaining dinner with a philosopher who told me the following story: One day in church he noticed that a priest, in a kneeling position, was hovering nine inches above the church floor. My natural scepticism of my dinner companion was increased when he went on to relate two further eye-witness experiences. He said that, among his many careers, he had once been warden of a home for delinquent boys, and he discovered that all the boys had 'I love my mummy' tattooed on their penises. An improbable story in itself, but not impossible. Unlike the case of the levitating priest, no great scientific principles would be called in question if it were true. Nevertheless, it seemed to provide a useful perspective on my neighbour's credibility. On another occasion, said this prolific raconteur, he had observed a crow strike a match while raising one wing to shield it from the wind. I forget whether the crow actually took a drag on a cigarette, but in any case the three stories, taken together, seemed to establish my companion as an unreliable, though diverting, witness. To put it mildly, the hypothesis that he was a liar (or a lunatic, or a hallucinating fantasist, or that he was researching the credulity of Oxford dons) seemed more probable than the alternative hypothesis that all three of his far-fetched stories were true.

As a philosopher, he would have known the logical test set out by the great eighteenth-century Scottish philosopher David Hume, which seems to me unassailable:

> . . . *no testimony is sufficient to establish a miracle, unless the testimony be of such a kind, that its falsehood would be more miraculous than the fact which it endeavours to establish.* 'Of Miracles' (1748)

I'll follow through Hume's meaning with respect to one of the best attested miracles of all time, one that, it is claimed, was witnessed by 70,000 people, and within living memory. This is the apparition of Our Lady of Fatima. I quote from an account in a Roman Catholic website which notes that, of the many claimed Marian sightings, this one is unusual in being officially recognized by the Vatican.

On October 13, 1917, there were more than 70,000 people gathered in the Cova da Iria in Fatima, Portugal. They had come to observe a miracle which had been foretold by the Blessed Virgin to three young visionaries: Lucia dos Santos, and her two cousins, Jacinta and Francisco Marto . . . Shortly after noon, Our Lady appeared to the three visionaries. As the Lady was about to leave, she pointed to the sun. Lucy excitedly repeated the gesture, and the people looked into the sky . . . Then a gasp of terror rose from the crowd, for the sun seemed to tear itself from the heavens and come crashing down upon the horrified multitude . . . Just when it seemed that the ball of fire would fall upon and destroy them, the miracle ceased, and the sun resumed its normal place in the sky, shining forth as peacefully as ever.

If the miracle of the moving sun had been seen only by Lucia, the young woman responsible for the cult of Fatima in the first place, not many would take it seriously. It could so easily be a private hallucination, or an obviously motivated lie. It is the 70,000 witnesses that impress. Could 70,000 people simultaneously be the victims of the same hallucination? Could 70,000 people collude in the same lie? Or if there never were 70,000 witnesses, could the reporter of the event get away with inventing so many?

Let's apply Hume's criterion. On the one hand, we are asked to believe in a mass hallucination, a trick of the light, or mass lie involving 70,000 people. This is admittedly improbable. But it is *less* improbable than the alternative: that the sun really did move. The sun hanging over Fatima was not, after all, a private sun; it was the same sun that warmed all the other millions of people on the daylight side of the planet. If the sun had moved in truth, but the event was seen only by the people of Fatima, an even greater miracle would have to have been perpetrated: an illusion of *non*-movement had to be staged for all the millions of witnesses not in Fatima. And that's ignoring the fact that, if the sun had really moved at the speed reported, the solar system would have broken up. We have no alternative but to follow Hume, choose the less miraculous of the available alternatives and conclude, contrary to official Vatican doctrine, that the miracle of Fatima never happened. Moreover, it is not at all clear that the onus is on us to explain how those 70,000 witnesses were misled.

Hume's is still an argument about the balance of probabilities. Moving to the far end of our spectrum of putative miracles, are there any speculations or allegations that we can utterly, and for all time, rule out? Physicists agree that if an inventor applies for a patent for a perpetual motion machine you can safely turn down his patent without even looking at his design. This is because any perpetual motion machine would violate the laws of thermodynamics. Sir Arthur Eddington wrote:

If someone points out to you that your pet theory of the universe is in disagreement with Maxwell's equations – then so much the worse for Maxwell's equations. If it is found to be contradicted by observation – well, these experimentalists do bungle things sometimes. But if your theory is found to be against the second law of thermodynamics I can give you no hope; there is nothing for it but to collapse in deepest humiliation. *The Nature of the Physical World* (1928)

Eddington is cleverly bending over backwards to make overwhelming concessions in the first part of the passage, so that his confidence in the second part has the more impact. But if you still find it too cocksure; if you think it is asking for trouble at the hands of some as yet unimaginable future technology, so be it. I won't press the point, but will take my weaker stand, with Hume, on relative probabilities. Fraud, illusion, trickery, hallucination, honest mistake or outright lies – the combination adds up to such a *probable* alternative that I shall always doubt *casual* observations or second-hand stories that seem to suggest the catastrophic overthrow of existing science. Existing science will undoubtedly be overthrown; not, however, by casual anecdotes or performances on television, but by rigorous research, repeated, dissected and repeated again.

Returning to our spectrum of improbabilities, fairies would fall somewhere between Apthorpe's aunt and a perpetual motion machine. If tiny, butterfly-sized humans, wearing wings and fashionable but miniature clothes, were authentically discovered tomorrow, no great principles of physics would have been violated. It wouldn't be nearly as revolutionary as a perpetual motion machine. On the other hand biologists would have a hard time fitting fairies into their existing classificatory scheme. Where did they spring from in evolution? Neither the fossil record nor existing zoology shows us any primates equipped with flapping wings, and it would be surprising indeed if they suddenly and uniquely evolved in a species sufficiently close to our own to have co-opted – as some famous fake photographs which excited the notoriously gullible Sir Arthur Conan Doyle clearly showed – 1920s-style clothes *à la mode*.

Alleged creatures such as the Loch Ness Monster, the Yeti or 'Abominable Snowman' of the Himalayas, and the dinosaur of the Congo, lie in the spectrum somewhere on the more probable side of Conan Doyle's fairies. There really is no particular reason why a relict population of plesiosaurs should not survive in Loch Ness. I can't tell you how delighted I, and all zoologists, would be if they did; or if an authentic dinosaur were found up the Congo. No biological and certainly no physical principles would be violated by such a discovery. The only reason it seems unlikely is that the last known dinosaur lived 65 million years ago, and 65 million years is a long time for a breeding population to remain concealed and unfossilized. As for the Yeti, the prospect of a surviving population of *Homo erectus*, or *Gigantopithecus*, would fill me with elation, if only I could believe it. I dearly wish I thought the idea more probable than the Humean alternatives – hallucinations, lying travellers' tales or honest misreadings of sun-enlarged animal footprints.

On 30 August 1938, Orson Welles's still famous radio dramatization of H. G. Wells's *The War of the Worlds* provoked widespread panic and even some rumoured suicides among listeners who thought its opening scene was – as it purported to be – an authentic news bulletin announcing a Martian invasion. This story is often held up as evidence of the laughable gullibility of the American nation; rather unfairly, I have always thought, for an invasion from outer space is not impossible and, were it to happen, a sudden newsflash on the radio is exactly how we'd probably first hear of it.

Flying saucer stories are perennially popular, but they tend to be disbelieved by the scientific community. Why? Not because a visitation from outer space is impossible or even wildly improbable. It is because, once again, the alternative explanations of fraud or illusion are more probable. As a matter of fact, numerous flying saucer stories have been painstakingly investigated, in wearisome detail, by teams of conscientious amateur and professional scientists. Time after time after time the stories have crumbled under investigation. Often they turn out to be straightforward hoaxes (lucrative for the perpetrators, because publishers pay good money for such stories, however poorly documented they may be, and whole industries of T-shirts and souvenir mugs can be supported). Or the 'saucers' turn out to have been aircraft, airships or balloons, seen, or illuminated, from a peculiar angle. Sometimes they are mirages or other tricks of the light, sometimes sightings of secret military aircraft.

One day, maybe, we shall be visited by extraterrestrial spaceships. But the odds that any *particular* report of flying saucers is genuine are low compared to the odds of the Humean alternatives of fraud or illusion. In particular, the thing that for me subtracts verisimilitude from most flying saucer stories is the almost comical resemblance of the reported aliens to ordinary humans, or to the latest fictional creations to have appeared on television. Many of them resemble human males sufficiently closely to want to copulate with human females, and even produce fertile offspring. As Carl Sagan and others have pointed out, abduction-crazed humanoid aliens seem to be the modern counterpart of seventeenth-century demons and witches.

Abetted by the prestige of television and the newspapers, astrology, paranormalism and alien visitations have a privileged inside track into the popular consciousness. If I am right that this tendency exploits our natural and laudable appetite for wonder, we have here paradoxical grounds for encouragement. We should take comfort from the thought that, since the appetite for wonder is fed so much more satisfyingly by real science, it ought to be a simple matter of education to combat superstition. But I suspect that there is an additional force at work which may make things more difficult. It is quite an interesting psychological force in its own right, and my purpose in the rest of this chapter is to explain it, because understanding it may help us to limit the damage it can cause. The additional force I am speaking of is a normal and, from many points of view, desirable credulity in children which, unless we are careful, can spill over into adulthood, with unfortunate results. I'll begin with a personal anecdote.

On All Fools' Day one year, when my sister and I were children, our parents and our uncle and aunt played a simple trick on us. They announced that they had rediscovered in the attic a little aeroplane which had belonged to them when young and they were going to take us up for a ride. Flying was less commonplace then, and we were thrilled. The only stipulation was that we had to be blindfolded. They led us by the hand, giggling and stumbling across the lawn, and strapped us into our seats. We heard the noise of the engine starting up, there was a lurch and up we went for a bumpy, swaying, reeling ride. From time to time we evidently passed through the high treetops, for we felt the branches gently brushing us and a pleasant, rushing wind in our faces. Finally we 'landed', the lurching ride came to an end on *terra firma*, the blindfold was removed and amid laughter all was revealed. There was no aeroplane. We had not travelled from the spot on the lawn where we had started. We had simply been sitting on a garden seat which our father and uncle had lifted and slewed and bumped around to simulate aerial movement. No engine, only the noisy vacuum cleaner, and a fan to blow wind in our faces. They and the tree branches brushing against us had been wielded by our mother and aunt standing by the seat. It had been fun while it lasted.

Credulous, faith-filled children that we were, we had looked forward to the promised flight for days before it happened. It never occurred to us to wonder why we must be blindfolded. Wouldn't it have been natural to ask what was the point of going for a joyride if you couldn't see anything? But no, our parents simply told us that, for some reason unspecified, it was necessary to blindfold us; and we accepted it. Perhaps they fell back on the time-honoured recipe of 'not spoiling the surprise'. We never wondered why our elders had kept from us the secret that at least one of them must be a trained pilot – I don't think we even asked which one. We just didn't have the sceptic's turn of mind. We had no fear of crashing, such was our faith in our parents. And when the blindfolds were removed and the joke was on us, we still didn't stop believing in Father Christmas, the tooth fairy, angels, heaven, the Happy Hunting Ground and the other stories that those same elders had told us. Incidentally, my mother has no memory of the incident, but she does remember the occasion in her own childhood when her father played the identical trick on herself and her little sister. His patter was even more far-fetched, because his plane 'took off' indoors and the children were told 'to duck as they flew out through the window'. She and her sister still fell for it.

Children are naturally credulous. Of course they are, what else would you expect? They arrive in the world knowing nothing, surrounded by adults who know, by comparison, everything. It is earnestly true that fire burns, that snakes bite, that if you walk unprotected in the noon sun you will bake red, raw and, as we now know, cancerous. Moreover, the other and apparently more scientific way to gain useful knowledge, learning by trial and error, is often a bad idea because the errors are too costly. If your mother tells you never to paddle in the lake because of the crocodiles, it is no good coming over all sceptical and scientific and 'adult' and saying, 'Thank you mother, but I prefer to put it to the experimental test.' Too often, such experiments would be terminal. It is easy to see why natural selection – the survival of the fittest – might penalize an experimental and sceptical turn of mind and favour simple credulity in children.

But this has an unfortunate by-product which can't be helped. If your parents tell you something that is not true, you must believe that, too. How could you not? Children are not equipped to know the difference between a true warning about genuine dangers and a false warning about going blind, say, or going to hell, if you 'sin'. If they were so equipped, they wouldn't need warnings at all. Credulity, as a survival device, comes as a package. You believe what you are told, the false with the true. Parents and elders know so much, it is natural to assume that they know everything and natural to believe them. So when they tell you about Father Christmas coming down the chimney, and about faith 'moving mountains', of course you believe that, too.

Children are gullible because they need to be if they are to fulfil their 'caterpillar' role in life. Butterflies have wings because their role is to locate members of the opposite sex and spread their offspring to new food plants. They have modest appetites satisfied by occasional sips of nectar. They eat little protein by comparison with caterpillars, which constitute the growing stage in the life history. Juvenile animals in general have the role of preparing to become successfully reproducing adults. Caterpillars are there to feed as rapidly as possible in order to chrysalize into flying, reproducing, dispersing adults. To this end they have no wings but instead have stout munching jaws and voracious, single-minded appetites.

Human children need to be credulous for a similar reason. They are information caterpillars. They are there to become reproducing adults, in a sophisticated, knowledge-based society. And by far the most important source of their information diet is their elders, above all their parents. For the same kind of reason as caterpillars have chumbling, hoovering jaws for sucking up cabbage flesh, human children have wide open ears and eyes, and gaping, trusting minds for sucking up language and other knowledge. They are suckers for adult knowledge. Tidal waves of data, gigabytes of wisdom flood through the portals of the infant skull, and most of it originates in the culture built up by parents and generations of ancestors. It is important, incidentally, not to take the caterpillar analogy too far. Children change gradually into adults, not suddenly, as caterpillars metamorphose into butterflies.

I remember once trying gently to amuse a six-year-old child at Christmas time by reckoning up with her how long it would take Father Christmas to go down all the chimneys in the world. If the average chimney is 20 feet long and there are, say, 100 million houses with children, how fast, I wondered aloud, would he have to whizz down each chimney in order to finish the job by dawn on Christmas Day? He'd hardly have time to tiptoe noiselessly into each child's bedroom, would he, since he'd necessarily be breaking the sound barrier? She saw the point and realized that there was a problem, but it didn't worry her in the least. She dropped the subject without pursuing it. The obvious possibility that her parents had been telling falsehoods never seemed to cross her mind. She wouldn't have put it in these words but the implication was that, if the laws of physics rendered Father Christmas's feat impossible, so much the worse for the laws of physics. It was enough that her parents had told her he went down all the chimneys during the few hours of Christmas Eve. It must be so because Mummy and Daddy said it was.

My contention is that trusting credulity may be normal and healthy in a child but it can become an unhealthy and reprehensible gullibility in an adult. Growing up, in the fullest sense of the word, should include the cultivation of a healthy scepticism. An active readiness to be deceived can be called childish because it is common – and defensible – among children. I suspect that its persistence in adults stems from a hankering after, indeed a pining for, the lost securities and comforts of childhood. The point was well put in 1986 by that great writer of popular science and science fiction Isaac Asimov: 'Inspect every piece of pseudoscience and you will find a security blanket, a thumb to suck, a skirt to hold.' Childhood is, for many people, a lost Arcadia, a kind of heaven, with its certainties and its securities, its fantasies of flying to the Never Never Land, its bedtime stories before we drifted off to the Land of Nod in the arms of Teddy Bear. With hindsight, the years of childish innocence may pass too soon. I love my parents for taking me for a ride, high as a kite, through the treetops; and for telling me about the Tooth Fairy and Father Christmas, about Merlin and his spells, about baby Jesus and the Three Wise Men. All these stories enrich childhood and, together with so many other things, help to make it, in memory, a time of enchantment.

The adult world may seem a cold and empty place, with no fairies and no Father Christmas, no Toyland or Narnia, no Happy Hunting Ground where mourned pets go, and no angels – guardian or garden variety. But there are also no devils, no hellfire, no wicked witches, no ghosts, no haunted houses, no dæmonic possession, no bogeymen or ogres. Yes, Teddy and Dolly turn out not to be really alive. But there are warm, live, speaking, thinking, adult bedfellows to hold, and many of us find it a more rewarding kind of love than the childish affection for stuffed toys, however soft and cuddly they may be.

Not to grow up properly is to retain our 'caterpillar' quality from childhood (where it is a virtue) into adulthood (where it becomes a vice). In childhood our credulity serves us well. It helps us to pack, with extraordinary rapidity, our skulls full of the wisdom of our parents and our ancestors. But if we don't grow out of it in the fullness of time, our caterpillar nature makes us a sitting target for astrologers, mediums, gurus, evangelists and quacks. The genius of the human child, mental caterpillar extraordinary, is for soaking up information and ideas, not for criticizing them. If critical faculties later grow it will be in spite of, not because of, the inclinations of childhood. The blotting paper of the child's brain is the unpromising seedbed, the base upon which later the sceptical attitude, like a struggling mustard plant, may possibly grow. We need to replace the automatic credulity of childhood with the constructive scepticism of adult science.

But I suspect an additional problem. Our story of the child as information caterpillar was too simple. The programming of the child's credulity has a twist which, until we understand it, is almost paradoxical. Let us go back to our picture of the child needing to absorb information from the previous generation as swiftly as possible. What if two adults, say your mother and your father, give you contradictory advice? What if your mother tells you that all snakes are deadly and you must never go near them, but next day your father tells you that all snakes are deadly except green ones and you can keep a green snake as a pet? Both pieces of advice may be good. The mother's more general advice has the desired effect of protecting you against snakes, even though it is too sweeping when it comes to green snakes. The father's more discriminating advice has the same protective effect and is in some ways better, but it could be fatal if carried, unrevised, to a distant country. In any case, to the young child the contradiction between the two might be dangerously confusing. Parents often make strenuous efforts not to contradict one another, and they are probably wise to do so. But natural selection, in 'designing' credulity, would need to build in a way of coping with contradictory advice. Perhaps a simple override rule, such as 'Believe whichever story you heard first.' Or 'Believe mother rather than father, and father rather than other adults in the population.'

Sometimes the advice from parents is specifically aimed against credulity towards other adults in the population. The following is a piece of advice that parents need to give their children: 'If any adult asks you to come with him and says that he is a friend of your parents, don't believe him, however nice he seems and even (or especially) if he offers you sweets. Only go with an adult that you and your parents already know, or who is wearing a policeman's uniform.' (A charming story recently appeared in the English newspapers in which Queen Elizabeth the Queen Mother, aged 97, told her chauffeur to stop the car when she noticed a crying child who was apparently lost. The kind old lady got out to comfort the little girl and offered to take her home. 'I can't,' wailed the child, 'I'm not allowed to talk to strangers.') A child is called upon to exercise the exact opposite of credulity in some circumstances: a tenacious persistence in believing an earlier adult statement in the face of what may be a tempting and plausible – but contradictory – later statement.

On their own, then, the words 'gullible' and 'credulous' are not quite right for children. Truly credulous people believe whatever they have most recently been told, even if this contradicts what others have told them before. The quality of childhood that I am trying to pin down is not pure gullibility but a complex combination of gullibility coupled with its opposite – stubborn persistence in a belief, once acquired. The full recipe, then, is extreme early gullibility followed by equally obstinate subsequent unshakeability. You can see what a devastating combination this could be. Those old Jesuits knew what they were about: 'Give me the child for his first seven years, and I'll give you the man.'

7

UNWEAVING THE UNCANNY

... though no great minist'ring reason sorts
Out the dark mysteries of human souls
To clear conceiving ...

JOHN KEATS 'Sleep and Poetry' (1817)

The eminent fertility specialist Robert Winston imagines the following advertisement, placed in the newspaper by an unscrupulous quack doctor, aimed at people who want their next baby to be, say, a son (the sexism underlying this assumption is not mine but could be found unquestioned all over the ancient world, and still in many places today). 'Send £500 for my patent recipe to make your baby a boy. Money refunded in full if I fail.' The money back guarantee is intended to establish confidence in the method. In fact, of course, since boys turn up anyway on approximately 50 per cent of occasions, the scheme would be a nice little earner. Indeed, the quack could safely offer compensation of, say, £250 for every girl born, over and above the money back guarantee. He would still show a tidy profit in the long run.

I used a similar illustration in one of my Royal Institution Christmas Lectures in 1991. I said I had reason to believe that among my audience was a psychic, clairvoyant individual, capable of influencing events purely by the power of thought. I would try to flush this individual out. 'Let's first establish,' I said, 'whether the psychic is in the left half or the right half of the lecture hall.' I invited everybody to stand up while my assistant tossed a coin.

Everybody on the left of the hall was asked to 'will' the coin to come down heads. Everybody on the right had to will it to be tails. Obviously one side had to lose, and they were asked to sit down. Then those that remained were divided into two, with half 'willing' heads and the other half tails. Again the losers sat down. And so on by successive halvings until, inevitably, after seven or eight tosses, one individual was left standing. 'A big round of applause for our psychic.' He must be psychic, mustn't he, because he successfully influenced the coin eight times in a row?

If the lectures had been televised live, instead of recorded and broadcast later, the demonstration would have been much more impressive. I'd have asked everybody who watched it whose surname begins before J in the alphabet to 'will' heads and the rest tails. Whichever half turned out to contain the 'psychic' would have been divided in half again, and so on. I'd have asked everybody to keep a written record of the order of their 'willings'. With two million viewers, it would have taken about 21 steps to narrow down to a single individual. To be on the safe side I'd have stopped a bit short of 21 steps. At, say, the eighteenth step I'd have invited anybody still in the game to phone in. There would have been quite a few and, with luck, one would phone. This individual would then have been invited to read out his/her written record: HTTTHHTHHHHTTTH HTT which would have matched the official record. So this one individual succeeded in influencing 18 successive tosses of a coin. Gasps of admiration. But admiration for what? Nothing but pure luck. I don't know if that experiment has been done. Actually, the trick here is so obvious it probably wouldn't fool many people. But how about the following?

A well-known 'psychic' goes on television, a lucrative engagement fixed up over lunch by his publicity agent. Staring out of ten million screens with hypnotically smouldering eyes (nice job by Make-up and Lighting), our imaginary seer intones that he feels a strange, spiritual rapport, a vibrating resonance of cosmic energy, with certain members of his audience. They will be able to tell who they are because, even as he utters his mystic incantation, *their watches will stop.* After only a brief pause, a telephone on his table rings and an amplified voice in awed tones announces that its owner's watch stopped dead within seconds of the clairvoyant's words. The caller adds that she had a premonition that this was going to happen even before she looked down at her watch, for something in her hero's burning eyes seemed to speak directly to her soul. She felt the 'vibrations' of 'energy'. Even as she is speaking, a second telephone rings. Yet another watch has stopped.

A third caller's grandfather clock stopped – surely a weightier feat than stopping a little watch whose delicate hairspring would naturally be more susceptible to psychic forces than the massive pendulum of the grandfather! Another viewer's watch actually stopped a little *before* the celebrated mystic made his pronouncement – is this not an even more impressive feat of psychic control? Yet another watch has been more impatiently susceptible to occult forces. It had stopped a whole day before, *at the very moment* when its owner looked at the famous mystic's photograph in the newspaper. The studio audience gasps its appreciation. This, surely, is psychic power beyond all scepticism, for it happened a *whole day early*! 'There are more things in heaven and earth, Horatio . . .'

What we need is less gasping and more thinking. This chapter is about how to take the sting out of coincidence by quietly sitting down and calculating the likelihood that it would have happened anyway. In the course of this, we shall discover that to disarm apparently uncanny coincidences is more interesting than gasping over them anyway.

Sometimes the calculation is easy. In a previous book I gave away the number of the combination lock on my bicycle. I felt safe in doing so because obviously my books would never be read by the kind of person who would steal a bicycle. Unfortunately somebody did steal it, and I now have a new lock with a new number, 4167. I find this number easy to remember. 41 is imprinted in my memory as the arbitrary code used to identify my clothes and shoes at boarding school. 67 is the age at which I am due to retire. Obviously there is no interesting coincidence here: whatever the number had been, I'd have searched my life for a mnemonic recipe and I'd have found it. But mark the sequel. On the day of writing this, I received from my Oxford college a letter saying:

Each person authorized to use the photocopiers is issued with a personal code number which permits access. Your new number is 4167.

My first thought was that I'd undoubtedly lose this piece of paper (I quickly lost its equivalent last year) and I must immediately think of a formula to fix it in my memory. Something similar to the mnemonic by which I remember my bicycle combination, perhaps? So I looked again at the number on the letter and, to borrow a neat line from Fred Hoyle's science fiction novel *The Black Cloud*, the figures on the piece of paper seemed to swell to a gigantic size.

4167

I didn't need a new mnemonic. The number was identical. I rushed to tell my wife of the amazing coincidence, but on more sober reflection I shouldn't have bothered.

The odds of this happening by chance alone are easily calculated. The first digit could have been anything from 0 to 9. So there is a one in 10 chance of getting a 4 and matching the bicycle lock. For each of these ten possibilities, the second digit could have been anything from 0 to 9, so again there is a one in 10 chance of matching the bike lock's second dial. The odds of matching the first two digits is therefore one in 100 and, following the logic through the other two digits, the odds of matching all four digits of the bicycle lock is one in 10,000. It is this large number that is our protection against theft.

The coincidence is impressive. But what should we conclude? Has something mysterious and providential been going on? Have guardian angels been at work behind the scenes? Have lucky stars swum into Uranus? No. There is no reason to suspect anything more than simple accident. The number of people in the world is so large compared with 10,000 that somebody, at this very moment, is bound to be experiencing a coincidence at least as startling as mine. It just happens that today was my day to notice such a coincidence. It isn't even an added coincidence that it happened to me on this particular day, while I was writing this chapter. I had in fact written the first draft of the chapter some weeks ago. I reopened it today, after the coincidence occurred, in order to insert this anecdote. I shall surely reopen it many times to revise and polish, and I shall not remove the references to 'today': they were accurate when written. This is another way in which we habitually inflate the impressiveness of coincidence in order to make a good story.

We can do a similar calculation for the television guru whose psychic miasma seemed to stop people's watches, but we'll have to use estimates rather than exact figures. Any given watch has a certain low probability of stopping at any moment. I don't know what this probability is, but here's the kind of way in which we could come to an estimate. If we take just digital watches, their battery typically runs out within a year. Approximately, then, a digital watch stops once per year. Presumably clockwork watches stop more often because people forget to wind them and presumably digital watches stop less often because people sometimes remember to renew the battery ahead of time. But both kinds of watches probably stop as often again because they develop faults of one kind or another. So, let our estimate be that any given watch is likely to stop about once a year. It doesn't matter too much how accurate our estimate is. The principle will remain.

If somebody's watch stopped three weeks after the spell was cast, even the most credulous would prefer to put it down to chance.

We need to decide how large a delay would have been judged by the audience as sufficiently simultaneous with the psychic's announcement to impress. About five minutes is certainly safe, especially since he can keep talking to each caller for a few minutes before the next call ceases to seem roughly simultaneous. There are about 100,000 five-minute periods in a year. The probability that any given watch, say mine, will stop in a designated five-minute period is about 1 in 100,000. Low odds, but there are 10 million people watching the show. If only half of them are wearing watches, we could expect about 25 of those watches to stop in any given minute. If only a quarter of these ring in to the studio, that is 6 calls, more than enough to dumbfound a naive audience. Especially when you add in the calls from people whose watches stopped the day before, people whose watches didn't stop but whose grandfather clocks did, people who died of heart attacks and their bereaved relatives phoned in to say that their 'ticker' gave out, and so on. This kind of coincidence is celebrated in the delightfully sentimental old song, 'Grandfather's Clock:'

> *Ninety years without slumbering,*
> *Tick, tock, tick, tock,*
> *His life seconds numbering,*
> *Tick, tock tick, tock,*
> *It stopped . . . short . . . never to go again*
> *When the old man died.*

Richard Feynman, in a 1963 lecture published posthumously in 1998, tells the story of how his first wife died at 9.22 in the evening and the clock in her room was later found to have stopped at exactly 9.22. There are those who would revel in the apparent mystery of this coincidence and feel that Feynman has taken away something precious when he gives us a simple, rational explanation of the mystery. The clock was old and erratic and was in the habit of stopping if tilted out of the horizontal. Feynman himself frequently repaired it. When Mrs Feynman died, the nurse's duty was to record the exact time of death. She moved over to the clock, but it was in dark shadow. In order to see it, she picked it up – and tilted its face towards the light . . . The clock stopped. Is Feynman really spoiling something beautiful when he tells us what is surely the true – and very simple – explanation? Not for my money. For me, he is affirming the elegance and beauty of an orderly universe in which clocks stop for reasons, not to titillate human sentimental fancy.

At this point, I want to invent a technical term, and I hope you'll forgive an acronym. PETWHAC stands for Population of Events That Would Have Appeared Coincidental. Population may seem an odd word, but it is the correct statistical term. I won't keep using capital letters because they stand so unattractively on the page. Somebody's watch stopping within ten seconds of the psychic's incantation obviously belongs within the petwhac, but so do many other events. Strictly speaking, the grandfather clock's stopping should not be included. The mystic did not claim that he could stop grandfather clocks. Yet when somebody's grandfather clock did stop, they immediately telephoned in because they were, if anything, even more impressed than they would have been if their watch had stopped. The odd misconception is fostered that the psychic is even *more* powerful since he didn't even bother to mention that he could stop grandfather clocks, too! Similarly, he said nothing about watches stopping the day before or grandfathers' tickers suffering cardiac arrests.

People feel that such unanticipated events belong in the petwhac. It looks to them as though occult forces must have been at work. But when you start to think like this, the petwhac becomes really quite large, and therein lies the catch. If your watch stopped exactly 24 hours earlier, you would not have to be unduly gullible to embrace this event within the petwhac. If somebody's watch stopped exactly seven minutes before the spell, this might impress some people because seven is an ancient mystic number. And the same presumably goes for seven hours, seven days . . . The larger the petwhac, the less we *ought* to be impressed by the coincidence when it comes. One of the devices of an effective trickster is to make people think exactly the opposite.

By the way, I deliberately chose a more impressive trick for my imaginary psychic than is actually done with watches on television. The more familiar feat is to *start* watches that have stopped. The television audience is invited to get up and fetch, out of drawers or attics, watches that have broken down, and hold them while the psychic performs some incantation or does some hypnotic eye work. What is really going on is that the warmth of the hand melts oil that has coagulated and this starts the watch ticking, if only briefly. Even if this happens in only a small proportion of cases, this proportion, multiplied by the large audience, will generate a satisfactory number of dumbfounded telephone calls. Actually, as Nicholas Humphrey explains in his admirable exposé of supernaturalism *Soul Searching* (1995), it has been demonstrated that more than 50 per cent of broken watches start, at least momentarily, if they are held in the hand.

Here's another example of a coincidence, where it is clear how to calculate the odds. We shall use it to go on and see how odds are sensitive to changing the petwhac. I once had a girlfriend who had the same birth date (though not in the same year) as my previous girlfriend. She told a friend of hers who believed in astrology, and the friend triumphantly asked how I could possibly justify my scepticism in the face of such overwhelming evidence that I had unwittingly been brought together with two successive women on the basis of their 'stars'. Once again, let's just think it through quietly. It is easy to calculate the odds that two people, chosen entirely at random, will have the same birthday. There are 365 days in the year. Whatever the birthday of the first person, the chance that the second will have the same birthday is 1 in 365 (forgetting leap years). If we pair people off in any particular way, such as taking the successive women friends of any one man, the odds that they will share their birthday are 1 in 365. If we take ten million men (less than the population of Tokyo or Mexico City), this apparently uncanny coincidence will have happened to more than 27,000 of them!

Now let's think about the petwhac and see how the apparent coincidence becomes less impressive as it swells. There are many other ways in which we could pair people off and still end up noticing an apparent coincidence. Two successive girlfriends with the same surname, although unrelated, for instance. Two business partners with the same birthday would also come within the petwhac; or two people with the same birthday sitting next to one another on an aeroplane. Yet, in a well-loaded Boeing 747, the odds are actually better than 50 per cent that at least one pair of neighbours will share a birthday. We don't usually notice this because we don't look over each other's shoulders as we fill in those tedious immigration forms. But if we did, somebody on most flights would go away muttering darkly about occult forces.

The birthday coincidence is famously phrased in a more dramatic way. If you have a roomful of only 23 people, mathematicians can prove that the odds are just greater than 50 per cent that two of them will share the same birthday. Two readers of an earlier draft of the book asked me to justify this astonishing statement. I have every sympathy with people who are phobic about mathematical formulas, so I'll spell out an approximation in words.

It's easier to calculate the odds that there is *not* a pair of shared birthdays in the room. Let's pretend that leap years don't exist, and suppose you and I are among the 23 people in a room. My birthday is 26 March. I don't know when yours is but, since there are 364 days that are not 26 March, the odds are 364/365 (0.997) that your birthday is not mine. But the pairing of you with me is only one of many pairings that we could imagine in our roomful of 23 people. We have to multiply 364/365 by itself for each pairing. How many pairings? A first guess is 23 × 23 (= 529), but this is clearly too many. It allows each person to be paired with himself, which is absurd: obviously and trivially, we each share a birthday with ourself! So we must at least subtract 23 from our preliminary list of possible pairings, which gives us $(23 \times 23) - 23 = 506$. And our first guess also counts you/me as separate from me/you, whereas obviously, if I share my birthday with you, you must share your birthday with me. In other words it counts each pairing twice. So we must halve our 506, giving 253 as the number of pairings that we must consider. It takes too long to work out by hand, but a computer (or a table of logarithms) will quickly lead you to the conclusion that 364/365 multiplied by itself 253 times yields a number very close to 0.5. This is the chance that there will *not* be any shared birthdays in the room.

So there's an approximately even chance that at least one pair of individuals in a committee of 23 will share a birthday. If you do the equivalent working for 30 people you'll find that the number of pairings, namely half of $((30 \times 30) - 30)$ comes to 435. And 364/365 multiplied by itself 435 times is approximately 0.30. So the chance of a couple of shared birthdays here is about 70 per cent. You can ensure a good income for yourself if you go to a rugby ground every Saturday and bet even money that among the 30 players out on the field are two who share the same birthday. Most people's intuition would encourage them to bet against such a coincidence. But they'd be wrong. It is this kind of intuitive error that in general bedevils our assessment of 'uncanny' coincidences.

Here's an actual coincidence where, although it is a little harder, we can make a stab at estimating the odds approximately. My wife once bought for her mother a beautiful antique watch with a pink face. When she got it home and peeled off the price label she was amazed to find, engraved on the back of the watch, her mother's own initials, M.A.B. Uncanny? Eerie? Spine-crawling? Arthur Koestler, the famous novelist, would have read much into it. So would C. G. Jung, the widely admired psychologist and inventor of the 'collective unconscious', who also believed that a bookcase or a knife might be induced by psychic forces to explode spontaneously with a loud report. My wife, who has more sense, merely thought the coincidence of initials remarkably convenient and sufficiently amusing to justify telling the story to me – and here I am now telling it to a wider audience.

So, what really are the odds against a coincidence of this magnitude? We can begin by calculating them in a naive way. There are 26 letters in the alphabet. If your mother has three initials and you find a watch engraved with three letters at random, the odds that the two will coincide is $1/26 \times 1/26 \times 1/26$, or one in 17,576. There are about 55 million people in Britain. If every one of them bought an antique engraved watch we'd expect more than 3,000 of them to gasp with amazement when they discovered that the watch already bore their mother's initials.

But the odds are actually better than this. Our naive calculation made the incorrect assumption that each letter has a probability of 1/26 of being somebody's initial. This is the *average* probability for the alphabet as a whole, but some letters, such as X and Z, have a smaller probability. Others, including M, A and B, are commoner: think how much more impressed we'd be if the coinciding initials had been X.Q.Z. We can improve our estimate of odds by sampling a telephone directory. Sampling is a respectable way of estimating something that we cannot count directly. The London directory is a good place to sample because it is large and London happens to be where my wife bought the watch and where her mother lived. The London telephone directory contains about 85,060 column inches, or about 1.34 column miles, of private citizens' names. Of these, about 8,110 column inches are devoted to the Bs. This means that about 9.5 per cent of Londoners have a surname beginning with B: much more frequent than the figure for an average letter: 1/26, or 3.8 per cent.

So, the probability that a randomly chosen Londoner would have a surname beginning with B is about 0.095 (= 9.5 per cent). What about the corresponding probabilities that the forenames will begin with M or A? It would take too long to count forename initials right through the telephone book, and there'd be no point since the telephone book is itself only a sample. The easiest thing to do is take a subsample where forename initials are conveniently arranged in alphabetical order. This is true of the listings *within* any one surname. I shall take the commonest surname in England – Smith – and look at what proportion of the Smiths are M. Smith and what proportion are A. Smith. It is a reasonable hope that this will be approximately representative of the probabilities of forename initials for Londoners generally. It turns out that there are rather more than 20 column yards of Smiths altogether. Of these, 0.073 of them (53.6 column inches) are M. Smiths. The A. Smiths fill 75.4 column inches, representing 0.102 of all the Smiths.

If you are a Londoner and you have three initials, therefore, the chances of your initials being M.A.B. in that order are approximately 0.102 × 0.073 × 0.095 or about 0.0007. Since the population of Britain is 55 million, this should mean that about 38,000 of them have the initials M.A.B., but only if everybody among those 55 million has three initials. Obviously not everybody does but, looking down the telephone directory again, it seems that at least a majority do. If we make the conservative assumption that only half of British people have three initials, that still means that more than 19,000 British people have identical initials to my wife's mother. Any one of them could have bought that watch and gasped with astonishment at the coincidence. Our calculation has shown that there is no reason to gasp.

Indeed, when we think harder about the petwhac, we find that we have even less right to be impressed. M.A.B. were the initial letters of my wife's mother's maiden name. Her married initials of M.A.W. would have seemed just as impressive had they been found on the watch. Surnames beginning with W are nearly as common in the telephone book as those beginning with B. This consideration approximately doubles the petwhac, by doubling the number of people in the country who would have been deemed, by a coincidence hunter, to have 'the same initials' as my wife's mother.

Moreover, if somebody bought a watch and found it to be engraved not with her mother's initials but with her own, she might consider it an even greater coincidence and more worthy to be embraced within the (ever-growing) petwhac.

The late Arthur Koestler, as I have already mentioned, was a great enthusiast of coincidences. Among the stories that he recounts in *The Roots of Coincidence* (1972) are several originally collected by his hero, the Austrian biologist Paul Kammerer (famous for publishing a faked experiment purportedly demonstrating the 'inheritance of acquired characteristics' in the midwife toad). Here is a typical Kammerer story quoted by Koestler:

On September 18, 1916, my wife, while waiting for her turn in the consulting rooms of Prof. Dr J. v. H., reads the magazine Die Kunst; *she is impressed by some reproductions of pictures by a painter named Schwalbach, and makes a mental note to remember his name because she would like to see the originals. At that moment the door opens and the receptionist calls out to the patients: 'Is Frau Schwalbach here? She is wanted on the telephone.'*

It probably isn't worth trying to estimate the odds against this coincidence, but we can at least write down some of the things that we'd need to know. 'At that moment the door opens' is a little vague. Did the door open one second after she made the mental note to look up Schwalbach's paintings or 20 minutes? How long could the interval have been, leaving her still impressed by the coincidence? The frequency of the name Schwalbach is obviously relevant: we'd be less impressed if it had been Schmidt or Strauss; more impressed if it had been Twistleton-Wykeham-Fiennes or Knatchbull-Huguesson. My local library doesn't have the Vienna telephone book, but a quick look in another large Germanic telephone directory, the Berlin one, yields half a dozen Schwalbachs: the name is not particularly common, therefore, and it is understandable that the lady was impressed. But we need to think further about the size of the petwhac. Similar coincidences could have happened to people in other doctors' waiting rooms; and in dentists' waiting rooms, government offices and so on; and not just in Vienna but anywhere else. The quantity to keep bearing in mind is the number of *opportunities* for coincidence that would have been thought, if they had occurred, just as remarkable as the one that actually did occur.

Now let's take another kind of coincidence, where it is even harder to know how to start calculating odds. Consider the often-quoted experience of dreaming of an old acquaintance for the first time in years and then getting a letter from him, out of the blue, the next day. Or of learning that he died in the night. Or of learning that he didn't die in the night but his father did. Or that his father didn't die but won the football pools. See how the petwhac grows out of control when we relax our vigilance?

Often, these coincidence stories are gathered together from a large field. The correspondence columns of popular newspapers contain letters sent in by individual readers who would not have written but for the amazing coincidence that had happened to them. In order to decide whether we should be impressed, we need to know the circulation figure for the newspaper. If it is 4 million, it would be surprising if we did *not* read daily of some stunning coincidence, since a coincidence only has to happen to one of the 4 million in order for us to have a good chance of being told about it in the paper. It is hard to calculate the probability of a particular coincidence happening to one person, say a long-forgotten old friend dying during the night we happen to dream about him. But whatever this probability is, it is surely far greater than one in 4 million.

So, there really is no reason for us to be impressed when we read in the newspaper of a coincidence that has happened to one of the readers, or to somebody, somewhere in the world. This argument against being impressed is entirely valid. Nevertheless, there may be something lurking here that still bothers us. You may be happy to agree that, from the point of view of a reader of a mass-circulation newspaper, we have no right to be impressed at a coincidence that happens to another of the millions of readers of the same newspaper who bothers to write in. But it is much harder to shake the feeling of spine-chilled awe when the coincidence happens to *you yourself*. This is not just personal bias. One can make a serious case for it. The feeling occurs to almost everybody I meet; if you ask anybody at random, there is a good chance that they will have at least one pretty uncanny story of coincidence to relate. On the face of it, this undermines the sceptic's point about newspaper stories having been culled from a millions-strong readership – a huge catchment of opportunity.

Actually it doesn't undermine it, for the following reason. Each one of us, though only a single person, none the less amounts to a very large population of *opportunities* for coincidence. Each ordinary day that you or I live through is an unbroken sequence of events, or incidents, any of which is potentially a *co*incidence. I am now looking at a picture on my wall of a deep-sea fish with a fascinatingly alien face. It is possible that, at this very moment, the telephone will ring and the caller will identify himself as a Mr Fish. I'm waiting . . .

The telephone didn't ring. My point is that, whatever you may be doing in any given minute of the day, there probably is some other event – a phone call, say – which, if it were to happen, would with hindsight be rated an eerie coincidence. There are so many minutes in every individual's lifetime that it would be quite surprising to find an individual who had *never* experienced a startling coincidence. During this particular minute, my thoughts have strayed to a schoolfellow called Haviland (I don't remember his Christian name, nor what he looked like) whom I haven't seen or thought of for 45 years. If, at this moment, an aeroplane manufactured by the de Haviland company were to fly past the window, I'd have a coincidence on my hands. In fact I have to report that no such plane has been forthcoming, but I have now moved on to think about something else, which gives yet another opportunity for coincidence. And so the opportunities for coincidence go on throughout the day and every day. But the negative occurrences, the failures to coincide, are not noticed and not reported.

Our propensity to see significance and pattern in coincidence, whether or not there is any real significance there, is part of a more general tendency to seek patterns. This tendency is laudable and useful. Many events and features in the world really are patterned in a non-random way and it is helpful to us, and to animals generally, to detect these patterns. The difficulty is to navigate between the Scylla of detecting apparent pattern when there isn't any, and the Charybdis of failing to detect pattern when there is. The science of statistics is quite largely concerned with steering this difficult course. But long before statistical methods were formalized, humans and indeed other animals were reasonably good intuitive statisticians. It is easy to make mistakes, however, in both directions.

Here are some true statistical patterns in nature which are not totally obvious, and which humans have not always known.

True pattern	***Reason difficult to detect***
Sexual intercourse is statistically followed by birth about 266 days later	The exact interval varies around the average of 266 days. Intercourse more often than not fails to result in conception. Intercourse is often frequent anyway, so it is not obvious that conception results from that rather than from, say, eating, which is also frequent.
Conception is relatively probable in the middle of a woman's cycle, and relatively improbable near menstruation	See above. In addition, women who don't menstruate don't conceive. This is a spurious correlation which gets in the way and even, to a naive mind, suggests the opposite of the truth.

Smoking causes lung cancer	Plenty of people who smoke don't get lung cancer. Many people get lung cancer who never smoked.
In a time of bubonic plague, proximity to rats, and especially their fleas, tends to lead to infection	Lots of rats and fleas around anyway. Rats and fleas are associated with so many other things, such as dirt and 'bad air', that it is hard to know which of the many correlated factors is the important one. i.e. again, there are spurious correlations that get in the way.

Now here are some false patterns which humans have mistakenly thought they detected.

False pattern	***Reason easy to be misled***
Droughts can be brought to an end by a rain dance (or human sacrifice, or sprinkling goats' blood on a ferret's kidneys, or whatever arbitrary custom the particular theology lays down)	Occasionally, rains do chance to follow upon a rain dance (etc.), and these rare lucky strikes lodge in the memory. When the rain dance, say, is not followed by rain, it is assumed that some detail went wrong with the ceremony, or that the gods are angry for some other reason: it is always easy enough to find a sufficiently plausible excuse.
Comets and other astronomical events portend crises in human affairs	See above. Also, it is in the interests of astrologers to foster the myth, just as it is no doubt in the interests of priests and witch-doctors to foster the myths about rain dances and ferrets' kidneys.

After a run of ill-luck, good luck becomes more likely	If bad luck persists, we assume that the run of bad luck hasn't ended yet, and we look forward all the more to its eventual end. If bad luck does not persist, the prophecy is seen as fulfilled. We subconsciously *define* a 'run' of bad luck in terms of its end. Therefore it obviously has to be followed by good luck.

We are not the only animals to seek statistical patterns of non-randomness in nature, and we are not the only animals to make mistakes of the kind that might be called superstitious. Both these facts are neatly demonstrated in the apparatus called the Skinner box, after the famous American psychologist B. F. Skinner. A Skinner box is a simple but versatile piece of equipment for studying the psychology of, usually, a rat or a pigeon. It is a box with a switch or switches let into one wall which the pigeon (say) can operate by pecking. There is also an electrically operated feeding (or other rewarding) apparatus. The two are connected in such a way that pecking by the pigeon has some influence on the feeding apparatus. In the simplest case, every time the pigeon pecks the key it gets food. Pigeons readily learn the task. So do rats and, in suitably enlarged and reinforced Skinner boxes, so do pigs.

We know that the causal link between key peck and food is provided by electrical apparatus, but the pigeon doesn't. As far as the pigeon is concerned, pecking a key might as well be a rain dance. Moreover, the link can be quite a weak, statistical one. The apparatus may be set up so that, far from every peck being rewarded, only one in 10 pecks is rewarded. This can mean literally every tenth peck. Or, with a different setting of the apparatus, it can mean that one in 10 pecks *on average* is rewarded, but on any particular occasion the exact number of pecks required is determined at random. Or there may be a clock which determines that one tenth of the time, on average, a peck will yield reward, but it is impossible to tell *which* tenth of the time. Pigeons and rats learn to press keys even when, one might think, you'd need to be quite a good statistician to detect the cause–effect relationship. They can be worked up to a schedule in which only a very small proportion of pecks is rewarded. Interestingly, habits learned when pecks are only occasionally rewarded are more durable than habits learned when all pecks are rewarded: the pigeon is less swiftly discouraged when the rewarding mechanism is switched off altogether. This makes intuitive sense if you think about it.

Pigeons and rats, then, are quite good statisticians, able to pick up slight, statistical laws of patterning in their world. Presumably this ability serves them in nature as well as in the Skinner box. Life out there is rich in pattern; the world is a big, complicated Skinner box. Actions by a wild animal are frequently followed by rewards or punishments or other important events. The relationship between cause and effect is frequently not absolute but statistical. If a curlew probes mud with its long, curved bill, there is a certain probability that it will strike a worm. The relationship between probe events and worm events is statistical but real. A whole school of research on animals has grown up around so-called Optimal Foraging Theory. Wild birds show quite sophisticated abilities to assess, statistically, the relative food-richness of different areas and they switch their time between the areas accordingly.

Back in the laboratory, Skinner founded a large school of research using Skinner boxes for all kinds of detailed purposes. Then, in 1948, he tried a brilliant variant on the standard technique. He completely severed the causal link between behaviour and reward. He set up the apparatus to 'reward' the pigeon from time to time *no matter what the bird did.* Now all that the birds actually needed to do was sit back and wait for the reward. But in fact this is not what they did. Instead, in six out of eight cases, they built up – exactly as though they were learning a rewarded habit – what Skinner called 'superstitious' behaviour. Precisely what this consisted of varied from pigeon to pigeon. One bird spun itself round like a top, two or three turns anticlockwise, between 'rewards'. Another bird repeatedly thrust its head towards one particular upper corner of the box. A third bird showed 'tossing' behaviour, as if lifting an invisible curtain with its head. Two birds independently developed the habit of rhythmic, side-to-side 'pendulum swinging' of the head and body. This last habit, incidentally, must have looked rather like the courtship dance of some birds of paradise. Skinner used the word superstition because the birds behaved as if they thought that their habitual movement had a causal influence on the reward mechanism, when actually it didn't. It was the pigeon equivalent of a rain dance.

A superstitious habit, once established, might persist for hours, long after the reward mechanism had been switched off. The habits did not, however, remain unchanged in form. They drifted, like the progressive improvisations of an organist. In one typical case the pigeon's superstitious habit began as a sharp movement of the head from the middle position towards the left. As time went by, the movement became more energetic. Eventually the whole body moved in the same direction and a step or two would be taken with the legs. After many hours of 'topographic drift', this leftward stepping movement became the predominant feature of the habit. The superstitious habits themselves may have been derived from the species' natural repertoire, but it is still fair to say that performing them in this context, and performing them repeatedly, is unnatural for pigeons.

Skinner's superstitious pigeons were behaving like statisticians, but statisticians who have got it wrong. They were alert to the possibility of links between events in their world, especially links between rewards that they wanted and actions that it was in their power to take. A habit, such as shoving the head up into the corner of the cage, began by chance. The bird just happened to do it at the moment before the reward mechanism was due to clunk into action. Understandably enough, the bird developed the tentative hypothesis that there was a link between the two events. So it shoved its head into the corner again. Sure enough, by the luck of Skinner's timing mechanism, the reward came again. If the bird had tried the experiment of *not* shoving its head into the corner, it would have found that the reward came anyway. But it would have needed to be a better and more sceptical statistician than many of us humans are in order to try this experiment.

Skinner makes the comparison with human gamblers developing little lucky 'tics' when playing cards. This kind of behaviour is also a familiar spectacle on bowling greens. Once the 'wood' (ball) has left the bowler's hand there is nothing more he can do to encourage it to move towards the 'jack' (target ball). Nevertheless, expert bowlers nearly always trot after their wood, often still in the stooped position, twisting and turning their bodies as if to impart desperate instructions to the now indifferent ball, and often speaking futile words of encouragement to it. A one-arm bandit in Las Vegas is nothing more nor less than a human Skinner box. 'Key-pecking' is represented not just by pulling the lever but also, of course, by putting money in the slot. It really is a fool's game because the odds are known to be stacked in favour of the casino – how else would the casino pay its huge electricity bills? Whether or not a given lever pull will deliver a jackpot is determined at random. It is a perfect recipe for superstitious habits. Sure enough, if you watch gambling addicts in Las Vegas you see movements highly reminiscent of Skinner's superstitious pigeons. Some talk to the machine. Others make funny signs to it with their fingers, or stroke it or pat it with their hands. They once patted it and won the jackpot and they've never forgotten it. I have watched computer addicts, impatient for a server to respond, behaving in a similar way, say, knocking the terminal with their knuckles.

My informant about Las Vegas has also made an informal study of London betting shops. She reports that one particular gambler habitually runs, after placing his bet, to a certain tile in the floor, where he stands on one leg while watching the race on the bookmaker's television. Presumably he once won while standing on this tile and conceived the notion that there was a causal link. Now, if somebody else stands on 'his' lucky tile (some other sportsmen do this deliberately, perhaps to try to hijack some of his 'luck' or just to annoy him) he dances around it, desperately trying to get a foot on the tile before the race ends. Other gamblers refuse to change their shirt, or to cut their hair, while they are 'on a lucky streak'. In contrast one Irish punter, who had a fine head of hair, shaved it completely bald in a desperate effort to change his luck. His hypothesis was that he was having rotten luck on the horses *and* he had lots of hair. Perhaps the two were connected somehow; perhaps these facts were all part of a meaningful pattern! Before we feel too superior, let us remember that large numbers of us were brought up to believe that Samson's fortunes changed utterly after Delilah cut off his hair.

How can we tell which apparent patterns are genuine, which random and meaningless? Methods exist, and they belong in the science of statistics and experimental design. I want to spend a little more time explaining a few of the principles, though not the details, of statistics. Statistics can largely be seen as the art of distinguishing pattern from randomness. Randomness *means* lack of pattern. There are various ways of explaining the ideas of randomness and pattern. Suppose I claim that I can tell girls' handwriting from boys'. If I am right, this would have to mean that there is a real pattern relating sex to handwriting. A sceptic might doubt this, agreeing that handwriting varies from person to person but denying that there is a sex-related pattern to this variation. How should you decide whether my claim, or the sceptic's, is right? It is no use just accepting my word for it. Like a superstitious Las Vegas gambler, I could easily have mistaken a lucky streak for a real, repeatable skill. In any case, you have every right to demand evidence. What evidence should satisfy you? The answer is evidence that is publicly recorded, and properly analysed.

The claim is, in any case, only a statistical claim. I do not maintain (in this hypothetical example – in reality I am not claiming anything) that I can infallibly judge the sex of the author of a given piece of handwriting. I claim only that among the great variation that exists among handwriting, some component of that variation correlates with sex. Therefore, even though I shall often make mistakes, if you give me, say, 100 samples of handwriting I should be able to sort them into boys and girls more accurately than could be achieved purely by guessing at random. It follows that, in order to assess my claim, you are going to have to calculate how likely it is that a given result could have been achieved by guessing at random. Once again, we have an exercise in calculating the odds of coincidence.

Before we get to the statistics, there are some precautions you need to take in designing the experiment. The pattern – the non-randomness we seek – is a pattern relating sex to handwriting. It is important not to confound the issue with extraneous variables. The handwriting samples that you give me should not, for instance, be personal letters. It would be too easy for me to guess the sex of the writer from the content of the letter rather than from the handwriting. Don't choose all the girls from one school and all the boys from another. The pupils from one school might share aspects of their handwriting, learning either from each other or from a teacher. These could result in real differences in handwriting, and they might even be interesting, but they could be representative of different schools, and only incidentally of different sexes. And don't ask the children to write out a passage from a favourite book. I should be influenced by a choice of *Black Beauty* or *Biggles* (readers whose childhood culture is different from mine will substitute examples of their own).

Obviously, it is important that the children should all be strangers to me, otherwise I'd recognize their individual writing and hence know their sex. When you hand me the papers they must not have the children's names on them, but you must have some means of keeping track of whose is which. Put secret codes on them for your own benefit, but be careful how you choose the codes. Don't put a green mark on the boys' papers and a yellow mark on the girls'. Admittedly, I won't know which is which, but I'll guess that yellow denotes one sex and green the other, and that would be a big help. It would be a good idea to give every paper a code number. But don't give the boys the numbers 1 to 10 and the girls 11 to 20; that would be just like the yellow and green marks all over again. So would giving the boys odd numbers and the girls even. Instead, give the papers *random* numbers and keep the crib list locked up where I cannot find it. These precautions are those named 'double blind' in the literature of medical trials.

Let's assume that all the proper double blind precautions have been taken, and that you have assembled 20 anonymous samples of handwriting, shuffled into random order. I go through the papers, sorting them into two piles for suspected boys and suspected girls. I may have some 'don't knows', but let's assume that you compel me to make the best guess I can in such cases. At the end of the experiment I have made two piles and you look through to see how accurate I have been.

Now the statistics. You'd expect me to guess right quite often even if I was guessing purely at random. But *how* often? If my claim to be able to sex handwriting is unjustified, my guessing rate should be no better than somebody tossing a coin. The question is whether my actual performance is sufficiently different from a coin-tosser's to be impressive. Here is how to set about answering the question.

Think about all *possible* ways in which I could have guessed the sex of the 20 writers. List them in order of impressiveness, beginning with all 20 correct and going down to completely random (all 20 exactly wrong is nearly as impressive as all 20 exactly right, because it shows that I can discriminate, even though I perversely reverse the sign). Then look at the *actual* way I sorted them and count up the percentage of all possible sortings that would have been as impressive as the actual one, or more. Here's how to think about all possible sortings. First, note that there is only one way of being

100 per cent right, and one way of being 100 per cent wrong, but there are lots of ways of being 50 per cent right. One could be right on the first paper, wrong on the second, wrong on the third, right on the fourth . . . There are somewhat fewer ways of being 60 per cent right. Fewer ways still of being 70 per cent right, and so on. The number of ways of making a single mistake is sufficiently few that we can write them all down. There were 20 scripts. The mistake could have been made on the first one, or on the second one, or on the third one . . . or on the twentieth one. That is, there are exactly 20 ways of making a single mistake. It is more tedious to write down all the ways of making two mistakes, but we can calculate how many ways there are, easily enough, and it comes to 190. It is harder still to count the ways of making three mistakes, but you can see that it could be done. And so on.

Suppose, in this hypothetical experiment, two mistakes is actually what I did make. We want to know how good my score was, on a spectrum of all possible ways of guessing. What we need to know is how many possible ways of choosing are as good as, or better than, my score. The number as good as my score is 190. The number better than my score is 20 (one mistake) plus 1 (no mistakes). So, the total number as good as or better than my score is 211. It is important to add in the ways of scoring *better* than my actual score because they properly belong in the petwhac, along with the 190 ways of scoring exactly as well as I did.

We have to set 211 against the total number of ways in which the 20 scripts could have been classified by penny-tossers. This is not difficult to calculate. The first script could have been boy or girl: that is two possibilities. The second script also could have been boy or girl. So, for each of the two possibilities for the first script, there were two possibilities for the second. That is $2 \times 2 = 4$ possibilities for the first two scripts. The possibilities for the first three scripts are $2 \times 2 \times 2 = 8$. And the possible ways of classifying all 20 scripts are $2 \times 2 \times 2$ . . . 20 times, or 2 to the power 20. This is a pretty big number, 1,048,576.

So, of all possible ways of guessing, the proportion of ways that are as good as or better than my actual score is 211 divided by 1,048,576, which is approximately 0.0002, or 0.02 per cent. To put it another way, if 10,000 people sorted the scripts entirely by tossing pennies, you'd expect only two of them to score as well as I actually did. This means that my score is pretty impressive and, if I performed as well as this, it would be strong evidence that boys and girls differ systematically in their handwriting. Let me repeat that this is all hypothetical. As far as I know, I have no such ability to sex handwriting. I should also add that, even if there was good evidence for a sex difference in handwriting, this would say nothing about whether the difference is innate or learned. The evidence, at least if it came from the kind of experiment just described, would be equally compatible with the idea that girls are systematically taught a different handwriting from boys – perhaps a more 'ladylike' and less 'assertive' fist.

We have just performed what is technically called a test of statistical significance. We reasoned from first principles, which made it rather tedious. In practice, research workers can call upon tables of probabilities and distributions that have been previously calculated. We therefore don't literally have to write down all possible ways in which things could have happened. But the underlying theory, the basis upon which the tables were calculated, depends, in essence, upon the same fundamental procedure. Take the events that *could* have been obtained and throw them down repeatedly at random. Look at the *actual* way the events occurred and measure how extreme it is, on the spectrum of all possible ways in which they could have been thrown down.

Notice that a test of statistical significance does not prove anything conclusively. It can't rule out luck as the generator of the result that we observe. The best it can do is place the observed result on a par with a specified amount of luck. In our particular hypothetical example, it was on a par with two out of 10,000 random guessers. When we say that an effect is statistically significant, we must always specify a so-called p-value. This is the probability that a purely random process would have generated a result at least as impressive as the actual result. A p-value of 2 in 10,000 is pretty impressive, but it is still possible that there is no genuine pattern there. The beauty of doing a proper statistical test is that we know *how* probable it is that there is no genuine pattern there.

Conventionally, scientists allow themselves to be swayed by p-values of 1 in 100, or even as high as 1 in 20: far less impressive than 2 in 10,000. What p-value you accept depends upon how important the result is, and upon what decisions might follow from it. If all you are trying to decide is whether it is worth repeating the experiment with a larger sample, a p-value of 0.05, or 1 in 20, is quite acceptable. Even though there is a 1 in 20 chance that your interesting result would have happened anyway by chance, not much is at stake: the error is not a costly one. If the decision is a life and death matter, as in some medical research, a much lower p-value than 1 in 20 should be sought. The same is true of experiments that purport to show highly controversial results, such as telepathy or 'paranormal' effects.

As we briefly saw in connection with DNA fingerprinting, statisticians distinguish false positive from false negative errors, sometimes called type 1 and type 2 errors respectively. A type 2 error, or false negative, is a failure to detect an effect when there really is one. A type 1 error, or false positive, is the opposite: concluding that there really is something going on when actually there is nothing but randomness. The p-value is the measure of the probability that you have made a type 1 error. Statistical judgement means steering a middle course between the two kinds of error. There is a type 3 error in which your mind goes totally blank whenever you try to remember which is which of type 1 and type 2. I still look them up after a lifetime of use. Where it matters, therefore, I shall use the more easily remembered names, false positive and false negative. I also, by the way, frequently make mistakes in arithmetic. In practice I should never dream of doing a statistical test from first principles as I did for the hypothetical handwriting case. I'd always look up in a table that somebody else – preferably a computer – had calculated.

Skinner's superstitious pigeons made false positive errors. There was in fact no pattern in their world that truly connected their actions to the deliveries of the reward mechanism. But they behaved as if they had detected such a pattern. One pigeon 'thought' (or behaved as if it thought) that left stepping caused the reward mechanism to deliver. Another 'thought' that thrusting its head into the corner had the same beneficial effect. Both were making false positive errors. A false negative error is made by a pigeon in a Skinner box who never notices that a peck at the key yields food if the red light is on, but that a peck when the blue light is on punishes by switching the mechanism off for ten minutes. There is a genuine pattern waiting to be detected in the little world of this Skinner box, but our hypothetical pigeon does not detect it. It pecks indiscriminately to both colours, and therefore gets a reward less frequently than it could.

A false positive error is made by a farmer who thinks that sacrificing to the gods brings longed-for rain. In fact, I presume (although I haven't investigated the matter experimentally), there is no such pattern in his world, but he does not discover this and persists in his useless and wasteful sacrifices. A false negative error is made by a farmer who fails to notice that there is a pattern in the world relating manuring of a field to the subsequent crop yield of that field. Good farmers steer a middle way between type 1 and type 2 errors.

It is my thesis that all animals, to a greater or lesser extent, behave as intuitive statisticians, choosing a middle course between type 1 and type 2 errors. Natural selection penalizes both type 1 and type 2 errors, but the penalties are not symmetrical and no doubt vary with the different ways of life of species. A stick caterpillar looks so like the twig it is sitting on that we cannot doubt that natural selection has shaped it to resemble a twig. Many caterpillars died to produce this beautiful result. They died because they did not sufficiently resemble a twig. Birds, or other predators, found them out. Even some very good twig mimics must have been found out. How else did natural selection push evolution towards the pitch of perfection that we see? But, equally, birds must many times have missed caterpillars because they resembled twigs, in some cases only slightly. Any prey animal, no matter how well camouflaged, can be detected by predators under ideal seeing conditions. Equally, any prey animal, no matter how poorly camouflaged, can be missed by predators under bad seeing conditions. Seeing conditions can vary with angle (a predator may spot a well-camouflaged animal when looking straight at it, but will miss a poorly camouflaged animal out of the corner of its eye). They can vary with light intensity (a prey may be overlooked at twilight, whereas it would be seen at noon). They can vary with distance (a prey which would be seen at six inches range may be overlooked at a range of 100 yards).

Imagine a bird cruising around a wood, looking for prey. It is surrounded by twigs, a very few of which might be edible caterpillars. The problem is to decide. We can assume that the bird could guarantee to tell whether an apparent twig was actually a caterpillar if it approached the twig really close and subjected it to a minute, concentrated examination in a good light. But there isn't time to do that for all twigs. Small birds with high turnover metabolism have to find food alarmingly often in order to stay alive. Any bird that scanned every individual twig with the equivalent of a magnifying glass would die of starvation before it found its first caterpillar. Efficient searching demands a faster, more cursory and rapid scanning, even though this carries a risk of missing some food. The bird has to strike a balance. Too cursory and it will never find anything. Too detailed and it will detect every caterpillar it looks at, but it will look at too few, and starve.

It is easy to apply the language of type 1 and type 2 errors. A false negative is committed by a bird that sails by a caterpillar without giving it a closer look. A false positive is committed by a bird that zooms in on a suspected caterpillar, only to discover that it is really a twig. The penalty for a false positive is the time and energy wasted flying in for the close inspection: not serious on any one occasion, but it could mount up fatally. The penalty for a false negative is missing a meal. No bird outside Cloud Cuckooland can hope to be free of all type 1 and type 2 errors. Individual birds will be programmed by natural selection to adopt some compromise policy calculated to achieve an optimum intermediate level of false positives and false negatives. Some birds may be biased towards type 1 errors, others towards the opposite extreme. There will be some intermediate setting which is best, and natural selection will steer evolution towards it.

Which intermediate setting is best will vary from species to species. In our example it will also depend upon conditions in the wood, for example, the size of the caterpillar population in relation to the number of twigs. These conditions may change from week to week. Or they may vary from wood to wood. Birds may be programmed to learn to adjust their policy as a result of their statistical experience. Whether they learn or not, successfully hunting animals must usually behave as if they are good statisticians. (I hope it is not necessary, by the way, to plod through the usual disclaimer: No, no, the birds aren't consciously working it out with calculator and probability tables. They are behaving *as if* they were calculating p-values. They are no more aware of what a p-value means than you are aware of the equation for a parabolic trajectory when you catch a cricket ball or baseball in the outfield.)

Angler fish take advantage of the gullibility of little fish such as gobies. But that is an unfairly value-laden way of putting it. It would be better not to speak of gullibility and say that they exploit the inevitable difficulty the little fish have in steering between type 1 and type 2 errors. The little fish themselves need to eat. What they eat varies, but it often includes small wriggling objects such as worms or shrimps. Their eyes and nervous systems are tuned to wriggling things. They look for wriggling movement and if they see it they pounce. The angler fish exploits this tendency. It has a long fishing rod, evolved from a modified spine, commandeered by natural selection from its original location at the front of the dorsal fin. The angler fish itself is highly camouflaged and it sits motionless on the sea bottom for hours at a time, blending perfectly with the weeds and rocks. The only part of it which is conspicuous is a 'bait', which looks like a worm, a shrimp or a small fish, at the end of its fishing rod. In some deep-sea species the bait is even luminous. In any case, it seems to wriggle like something worth eating when the angler waves its rod. A possible prey fish say, a goby, is attracted. The angler 'plays' its prey for a little while to hook its attention, then casts the bait down into the still unsuspected region in front of its own invisible mouth, and the little fish often follows. Suddenly that huge mouth is invisible no longer. It gapes massively, there is a violent inrushing of water, engulfing every floating object in the vicinity, and the little fish has pursued its last worm.

From the point of view of a hunting goby, any worm may be overlooked or it may be seen. Once the 'worm' has been detected, it may turn out to be a real worm or an angler fish's lure, and the unfortunate fish is faced with a dilemma. A false negative error would be to refrain from attacking a perfectly good worm for fear that it might be an angler fish lure. A false positive error would be to attack a worm, only to discover that it is really a lure. Once again, it is impracticable in the real world to get it right all the time. A fish that is too risk-averse will starve because it never attacks worms. A fish that is too foolhardy won't starve but it may be eaten. The optimum in this case may not be halfway between. More surprisingly, the optimum may be one of the extremes. It is possible that angler fish are sufficiently rare that natural selection favours the extreme policy of attacking all apparent worms. I am fond of a remark of the philosopher and psychologist William James on human angling:

There are more worms unattached to hooks than impaled upon them; therefore, on the whole, says Nature to her fishy children, bite at every worm and take your chances. (1910)

Like all other animals, and even plants, humans can and must behave as intuitive statisticians. The difference with us is that we can do our calculations twice over. The first time intuitively, as though we were birds or fish. And then again explicitly, with pencil and paper or computer. It is tempting to say that the pencil and paper way gets the right answer, so long as we don't make some publicly detectable blunder like adding in the date, whereas the intuitive way may yield the wrong answer. But there strictly is no 'right' answer, even in the case of pencil and paper statistics. There may be a right way to do the sums, to calculate the p-value, but the criterion, or threshold p-value, that we demand before choosing a particular action is still our decision and it depends upon our aversion to risk. If the penalty for making a false positive error is much greater than the penalty for making a false negative error, we should adopt a cautious, conservative threshold: almost never try a 'worm' for fear of the consequences. Conversely, if the risk-asymmetry is opposite, we should rush in and try every 'worm' that is going: it is unlikely to matter if we keep tasting false worms so we may as well have a go.

Taking on board the need to steer between false positive and false negative errors, let me return to uncanny coincidence and the calculation of the probability that it would have happened anyway. If I dream of a long-forgotten friend who dies the same night, I am tempted, like anyone else, to see meaning or pattern in the coincidence. I really have to force myself to remember that quite a few people die every night, masses of people dream every night, they quite often dream that people die, and coincidences like this are probably happening to several hundred people in the world every night. Even as I think this through, my own intuition cries out that there must be meaning in the coincidence because it has happened to *me*. If it is true that intuition is, in this case, making a false positive error, we need to come up with a satisfactory explanation for why human intuition errs in this direction. As Darwinians, we should be alive to the possible pressures towards erring on the type 1 or the type 2 side of the divide.

As a Darwinian, I want to suggest that our willingness to be impressed at apparently uncanny coincidence (which is a case of our willingness to see pattern where there is none) is related to the typical population size of our ancestors and the relative poverty of their everyday experience. Anthropology, fossil evidence and the study of other apes all suggest that our ancestors, for much of the past few million years, probably lived in either small roving bands or small villages. Either of these would mean that the number of friends and acquaintances that our ancestors would ordinarily meet and talk to with any frequency was not more than a few dozen. A prehistoric villager could expect to hear stories of startling coincidence in proportion to this small number of acquaintances. If the coincidence happened to somebody not in his village, he wouldn't hear the story. So our brains became calibrated to detect pattern and gasp with astonishment at a level of coincidence which would actually be quite modest if our catchment area of friends and acquaintances had been large.

Nowadays, our catchment area is large, especially because of newspapers, radio and other vehicles of mass news circulation. I've already spelled out the argument. The very best and most spine-creeping coincidences have the opportunity to circulate, in the form of bated-breath stories, over a far wider audience than was ever possible in ancestral times. But, I am now conjecturing, our brains are calibrated by ancestral natural selection to expect a much more modest level of coincidence, calibrated under small village conditions. So we are impressed by coincidences because of a miscalibrated gasp threshold. Our subjective petwhacs have been calibrated by natural selection in small villages, and, as is the case with so much of modern life, the calibration is now out of date. (A similar argument could be used to explain why we are so hysterically risk-averse to hazards that are much publicized in the newspapers – perhaps anxious parents who imagine ravening paedophiles lurking behind every lamp post on their children's walk from school are 'miscalibrated'.)

I guess that there may be another, particular effect pushing in the same direction. I suspect that our individual lives under modern conditions are richer in experiences per hour than were ancestral lives. We don't just get up in the morning, scratch a living in the same way as yesterday, eat a meal or two and go to sleep again. We read books and magazines, we watch television, we travel at high speed to new places, we pass thousands of people in the street as we walk to work. The number of faces we see, the number of different situations we are exposed to, the number of separate things that happen to us, is much greater than for our village ancestors. This means that the number of *opportunities* for coincidence is greater for each one of us than it would have been for our ancestors, and consequently greater than our brains are calibrated to assess. This is an additional effect, over and above the population size effect that I have already noted.

With respect to both these effects, it is theoretically possible for us to recalibrate ourselves, learn to adjust our gasp threshold to a level more appropriate to modern populations and modern richnesses of experience. But this seems to be revealingly difficult even for sophisticated scientists and mathematicians. The fact that we still do gasp when we do, that clairvoyants and mediums and psychics and astrologers manage to make such a nice living out of us, all suggests that we do not, on the whole, learn to recalibrate ourselves. It suggests that the parts of our brains responsible for doing intuitive statistics are still back in the stone age.

The same may be true of intuition generally. In *The Unnatural Nature of Science* (1992), the distinguished embryologist Lewis Wolpert has argued that science is difficult because it is more or less systematically counter-intuitive. This is contrary to the view of T. H. Huxley (Darwin's Bulldog) who saw science as 'nothing but trained and organized common sense, differing from the latter only as a veteran may differ from a raw recruit'. For Huxley, the methods of science 'differ from those of common sense only as far as the guardsman's cut and thrust differ from the manner in which a savage wields his club'. Wolpert insists that science is deeply paradoxical and surprising, an affront to common sense rather than an extension of it, and he makes a good case. For example, every time you drink a glass of water you are imbibing at least one molecule that passed through the bladder of Oliver Cromwell. This follows by extrapolation from Wolpert's observation that 'there are many more molecules in a glass of water than there are glasses of water in the sea'. Newton's law that objects stay in motion unless positively stopped is counter-intuitive. So is Galileo's discovery that, when there is no air resistance, light objects fall at the same rate as heavy objects. So is the fact that solid matter, even a hard diamond, consists almost entirely of empty space. Steven Pinker gives an illuminating discussion of the evolutionary origins of our physical intuitions in *How the Mind Works* (1998).

More profoundly difficult are the conclusions of quantum theory, overwhelmingly supported by experimental evidence to a stupefyingly convincing number of decimal places, yet so alien to the evolved human mind that even professional physicists don't understand them in their intuitive thoughts. It seems to be not just our intuitive statistics but our very minds themselves that are back in the stone age.

8

HUGE CLOUDY SYMBOLS OF A HIGH ROMANCE

To gild refined gold, to paint the lily,
To throw a perfume on the violet,
To smooth the ice, or add another hue
Unto the rainbow, or with taper-light
To seek the beauteous eye of heaven to garnish,
Is wasteful and ridiculous excess.

WILLIAM SHAKESPEARE,
King John, Act IV, scene ii

It is a central tenet of this book that science, at its best, should leave room for poetry. It should note helpful analogies and metaphors that stimulate the imagination, conjure in the mind images and allusions that go beyond the needs of straightforward understanding. But there's bad poetry as well as good, and bad poetic science can lead the imagination along false trails. That danger is the subject of this chapter. By bad poetic science I mean something other than incompetent or graceless writing. I am talking about almost the opposite: about the power of poetic imagery and metaphor to inspire bad science, even if it is good poetry, perhaps especially if it is good poetry, for that gives it the greater power to mislead.

Bad poetry, in the form of an over-indulgent eye for poetic allegory, or the inflation of casual and meaningless resemblances into huge cloudy symbols of a high romance (Keats's phrase), lurks behind many magical and religious customs. Sir James Frazer, in *The Golden Bough* (1922), recognizes a major category of magic which he calls homeopathic or imitative magic. The imitation varies from the literal to the symbolic. The Dyaks of Sarawak would eat the hands and knees of the slain in order to steady their own hands and strengthen their own knees. The bad poetic idea here is the notion that there is some essence of hand or essence of knee which can be transmitted from person to person. Frazer notes that, before the Spanish conquest, the Aztecs of Mexico

believed that by consecrating bread their priests could turn it into the very body of their god, so that all who thereupon partook of the consecrated bread entered into a mystic communion with the deity by receiving a portion of his divine substance into themselves. The doctrine of transubstantiation, or the magical conversion of bread into flesh, was also familiar to the Aryans of ancient India long before the spread and even the rise of Christianity.

Frazer later generalizes the theme:

It is now easy to understand why a savage should desire to partake of the flesh of an animal or man whom he regards as divine. By eating the body of the god he shares in the god's attributes and when he is a vine-god the juice of the grape is his blood; and so by eating the bread and drinking the wine the worshipper partakes of the real body and blood of his god. Thus the drinking of wine in the rites of a vine-god like Dionysus is not an act of revelry, it is a solemn sacrament.

All over the world, ceremonies are based upon an obsession with things *representing* other things that they slightly resemble, or resemble in one respect. Powdered rhinoceros horn is, with tragic consequences, thought to be aphrodisiac, apparently for no better reason than the superficial resemblance of the horn itself to an erect penis. To take another common practice, professional rainmakers frequently imitate thunder or lightning, or they conjure a miniature 'homeopathic dose' of rain by sprinkling water from a bundle of twigs. Such rituals can become elaborate and costly in time and effort.

Among the Dieri of central Australia, rainmaking wizards, symbolically *representative* of ancestor gods, were bled (dripping blood *represents* the longed-for rain) into a large hole inside a hut especially built for the purpose. Two rocks, intended to *stand for* clouds and presage rain, were then carried by the two wizards some 10 or 15 miles away, where they were placed atop a tall tree, to *symbolize* the height of the clouds. Meanwhile, back at the hut, the men of the tribe would stoop low and, without using their hands, charge at the walls and butt their way through with their heads. They continued butting back and forth until the hut was destroyed. The piercing of the walls with their heads *symbolized* the piercing of the clouds and, they believed, released rain from real clouds. As an added precaution, the Great Council of the Dieri would also keep a stockpile of boys' foreskins in constant readiness, because of their homeopathic power to produce rain (do penises not 'rain' urine – surely eloquent evidence of their power?).

Another homeopathic theme is the 'scapegoat' (so-called because a particular Jewish version of the rite involved a goat), in which a victim is chosen to embody, signify, or be loaded up with, all the sins and misfortunes of the village. The scapegoat is then driven out, or in some cases killed, carrying the evils of the people with him. Among the Garos people of Assam, near the foothills of the eastern Himalayas, a langur monkey (or sometimes a bamboo rat) used to be captured, led to every house in the village to soak up their evil spirits and then crucified on a bamboo scaffold. In Frazer's words, the monkey

> *is the public scapegoat, which by its vicarious sufferings and death relieves the people from all sickness and mishap in the coming year.*

In many cultures the scapegoat is a human victim, and often he is identified with a god. The symbolic notion of water 'washing' away sins is another common theme, sometimes combined with the idea of the scapegoat. In one New Zealand tribe,

a service was performed over an individual, by which all the sins of the tribe were supposed to be transferred to him, a fern stalk was previously tied to his person, with which he jumped into the river, and there unbinding, allowed it to float away to the sea, bearing their sins with it.

Frazer also reports that water was used by the rajah of Manipur as a vehicle to transfer his sins to a human scapegoat, who crouched under a scaffold on which the rajah took his bath, dripping water (and washed-away sins) on to the scapegoat below.

Condescension towards 'primitive' cultures is not admirable, so I have carefully chosen examples to remind us that theologies closer to home are not immune to homeopathic or imitative magic. The water of baptism 'washes' away sins. Jesus himself is a stand-in for humanity (in some versions via a symbolic standing in for Adam) in his crucifixion, which homeopathically atones for our sins. Whole schools of Mariology discern a symbolic virtue in the 'feminine principle'.

Sophisticated theologians who do not literally believe in the Virgin Birth, the Six Day Creation, the Miracles, the Transubstantiation or the Easter Resurrection are nevertheless fond of dreaming up what these events might symbolically *mean*. It is as if the double helix model of DNA were one day to be disproved and scientists, instead of accepting that they had simply got it wrong, sought desperately for a symbolic meaning so deep as to transcend mere factual refutation. 'Of course,' one can hear them saying, 'we don't literally believe *factually* in the double helix any more. That would indeed be crudely simplistic. It was a story that was right for its own time, but we've moved on. Today, the double helix has a new meaning for us. The compatibility of guanine with cytosine, the glove-like fit of adenine with thymine, and especially the intimate mutual twining of the left spiral around the right, all *speak to us* of loving, caring, nurturing relationships . . .' Well, I'd be surprised if it quite came to that, and not only because the double helix model is now very unlikely to be disproved. But in science, as in any other field, there really are dangers of becoming intoxicated by symbolism, by meaningless resemblances, and led farther and farther from the truth, rather than towards it. Steven Pinker reports that he is troubled by correspondents who have discovered that everything in the universe comes in threes:

the Father, the Son, and the Holy Ghost; protons, neutrons and electrons; masculine, feminine and neuter; Huey, Dewey, and Louie; and so on, for page after page. *How the Mind Works* (1998)

Slightly more seriously, Sir Peter Medawar, the distinguished British zoologist and polymath whom I quoted before invents a

great new universal principle of complementarity (not Bohr's) according to which there is an essential inner similarity in the relationships that hold between antigen and antibody, male and female, electropositive and electronegative, thesis and antithesis, and so on. These pairs have indeed a certain 'matching oppositeness' in common, but that is all *they have in common. The similarity between them is not the taxonomic key to some other, deeper affinity, and our recognizing its existence marks the end, not the inauguration, of a train of thought.*

Pluto's Republic (1982)

While I am quoting Medawar in the context of becoming intoxicated by symbolism, I cannot resist mentioning his devastating review of *The Phenomenon of Man* (1959), in which Teilhard de Chardin 'resorts to that tipsy, euphoristic prose poetry which is one of the more tiresome manifestations of the French spirit'. This book is, for Medawar (and for me now, although I confess that I was captivated when I read it as an over-romantic undergraduate), the quintessence of bad poetic science. One of the topics Teilhard covers is the evolution of consciousness, and Medawar quotes him as follows, again in *Pluto's Republic*:

By the end of the Tertiary era, the psychical temperature in the cellular world had been rising for more than 500 million years ... When the anthropoid, so to speak, had been brought 'mentally' to boiling-point some further calories were added ... No more was needed for the whole inner equilibrium to be upset ... By a tiny 'tangential' increase, the 'radial' was turned back on itself and so to speak took an infinite leap forward. Outwardly, almost nothing in the organs had changed. But in depth, a great revolution had taken place; consciousness was now leaping and boiling in a space of super-sensory relationships and representations ...

Medawar drily comments:

The analogy, it should be explained, is with the vaporization of water when it is brought to boiling-point, and the image of hot vapour remains when all else is forgotten.

Medawar also calls attention to the notorious fondness of mystics for 'energy' and 'vibrations', technical terms misused to create the illusion of scientific content where there is no content of any kind. Astrologers, too, think that each planet exudes its own, qualitatively distinct 'energy', which affects human life and has affinities with some human emotion: love in the case of Venus, aggression for Mars, intelligence for Mercury. These planetary qualities are based on – what else? – the characters of the Roman gods after whom the planets are named. In a style reminiscent of the aboriginal rainmakers, the Zodiacal signs are further identified with the four alchemical 'elements': earth, air, fire and water. People born under earth signs like Taurus are, to quote an astrological page chosen at random from the worldwide web, *dependable, realistic, down to earth ... People with water in their chart are sympathetic, compassionate, nurturing, sensitive, psychic, mysterious and possess an intuitive awareness ... Those who lack water may be unsympathetic and cold.*

Pisces is a water sign (I wonder why) and the element of water 'represents unconscious force's energy and power motivating us . . .'

Though Teilhard's book purports to be a work of science, his psychical 'temperature' and 'calories' seem approximately as meaningless as astrological planetary energies. The metaphorical usages are not usefully connected to their real-world equivalents. There is either no resemblance at all, or what resemblance there is impedes understanding rather than aids it.

With all this negativity, we mustn't forget that it is precisely the use of symbolic intuition to uncover genuine patterns of resemblance that leads scientists to their greatest contributions. Thomas Hobbes went too far when he concluded, in chapter 5 of *Leviathan* (1651), that

Reason *is the* pace; *Encrease of* Science, *the* way; *and the Benefit of man-kind, the* end. *And, on the contrary, Metaphors, and senselesse and ambiguous words, are like* ignes fatui; *and reasoning upon them, is wandering amongst innumerable absurdities; and their end, contention, and sedition, or contempt.*

Skill in wielding metaphors and symbols is one of the hallmarks of scientific genius.

The literary scholar, theologian and children's author C. S. Lewis, in a 1939 essay, made a distinction between magisterial poetry (in which scientists, say, use metaphoric and poetic language to explain to the rest of us something that we already understand) and pupillary poetry (in which scientists use poetic imagery to assist themselves in their own thinking). Important as both are, it is the second usage that I am emphasizing here. Michael Faraday's invention of magnetic 'lines of force', which we can think of as made of springy materials under tension, eager to release their energy (in the sense carefully defined by physicists) was vital to his own understanding of electromagnetism. I've already made use of the physicist's poetic image of inanimate entities – electrons, say, or light waves – striving to minimize their travel time. This is an easy way to get the right answer, and it is surprising how far it can be taken. I once heard Jacques Monod, the great French molecular biologist, say that he gained chemical insight by imagining how it would feel to be an electron at a particular molecular juncture. The German organic chemist Kekulé reported that he dreamed of the benzene ring in the form of a snake devouring its tail. Einstein was forever imagining: his extraordinary mind led by poetic thought-experiments through seas of thought stranger than even Newton voyaged.

But this chapter is about bad poetic science and we come down with a bump in the following example, sent me by a correspondent:

I consider our cosmic environment has a tremendous influence on the course of evolution. How else do we account for the helical structure of DNA which may be either due to the helical path of incoming solar radiation or the path of Earth orbiting the Sun which, due to its magnetic axis, tilted at 23.5° from the perpendicular, is helical, hence the solstices and equinoxes?

Realistically, there is not the smallest connection between the helical structure of DNA and the helical path of radiation or the planet's orbit. The association is superficial and meaningless. None of the three assists our understanding of any of the others. The author is drunk on metaphor, captivated by the idea of the helix, which misleads him into seeing connections which do not illuminate the truth in any way. Calling it poetic science is too kind: it is more like theological science.

Recently my incoming mail has registered a sharp rise in the normal load of 'chaos theory', 'complexity theory', 'non-linear criticality' and similar phrases. Now I'm not saying that these correspondents lack the faintest, foggiest clue what they are talking about. But I will say it's hard to discover whether they do. New Age cults of all kinds are swimming in bogus scientific language, regurgitated, half-understood (no, less than half) jargon: energy fields, vibration, chaos theory, catastrophe theory, quantum consciousness. Michael Shermer, in *Why People Believe Weird Things* (1997), quotes a typical example:

This planet has been slumbering for eons and with the inception of higher energy frequencies is about to awaken in terms of consciousness and spirituality. Masters of limitation and masters of divination use the same creative force to manifest their realities, however, one moves in a downward spiral and the latter moves in an upward spiral, each increasing the resonant vibration inherent in them.

Quantum uncertainty and chaos theory have had deplorable effects upon popular culture, much to the annoyance of genuine aficionados. Both are regularly exploited by those with a bent for abusing science and shanghaiing its wonder. They range from professional quacks to daffy New Agers. In America, the self-help 'healing' industry coins millions – and it has not been slow to cash in on quantum theory's formidable talent to bewilder. This has been documented by the American physicist Victor Stenger, author of the excellent *Physics and Psychics* (1990). One well-heeled healer wrote a string of bestselling books on what he calls 'quantum healing'. Another book in my possession has sections on quantum psychology, quantum responsibility, quantum morality, quantum aesthetics, quantum immortality and quantum theology. One feels vaguely let down that there is no 'quantum caring', but perhaps I missed it.

My next example packs a great deal of bad poetic science into a small space. It comes from the jacket blurb of a book:

A masterly description of the evolving, musical, nurturing and essentially caring universe.

Even if 'caring' were not a limp cliché, universes aren't the sort of entities to which a word like caring can sensibly be applied. (I realize that I am vulnerable to the criticism that a gene is not the sort of entity to which a word like 'selfish' should be applied. But I vigorously challenge anyone to maintain the criticism after reading *The Selfish Gene* itself, as opposed to just the title.) To apply 'evolving' to the universe is defensible but, as we shall see, it is probably best not to do so. 'Musical' is presumably an allusion to the Pythagorean 'music of the spheres', a piece of poetic science which may not have been bad originally but which we should have grown out of by now. 'Nurturing' has the smell of one of the most deplorable schools of bad poetic science, inspired by a misguided variant of feminism.

Here's another example. A number of scientists were invited by an anthologist in 1997 to send in the one question that they most wanted to see answered. Most of the questions were interesting and stimulating, but the following submission from one (male) individual is so absurd that I can only blame it on sucking up to feminist bullies:

What will happen when the male, scientific, hierarchical, control-oriented Western culture that has dominated Western thought integrates with the emerging female, spiritual, holographic, relationship-oriented Eastern way of seeing?

Did he mean 'holographic' or 'holistic'? Perhaps both. Who cares as long as it sounds good? Meaning is not what this is about.

The historian and philosopher of science Noretta Koertge, in her 1995 essay in *Skeptical Inquirer*, accurately puts her finger on the dangers of a kind of perverted feminism which could have a malign influence upon women's education:

Instead of exhorting young women to prepare for a variety of technical subjects by studying science, logic, and mathematics, Women's Studies students are now being taught that logic is a tool of domination ... the standard norms and methods of scientific inquiry are sexist because they are incompatible with 'women's ways of knowing.' The authors of the prize-winning book with this title report that the majority of the women they interviewed fell into the category of 'subjective knowers,' characterized by a 'passionate rejection of science and scientists.' These 'subjectivist' women see the methods of logic, analysis and abstraction as 'alien territory belonging to men' and 'value intuition as a safer and more fruitful approach to truth.'

One might have thought that, however dippy it might be, this kind of thinking would at least be gentle and, well, 'nurturing'. But the opposite is often true. At times it develops an ugly, hectoring tone, masculine in the worst sense. Barbara Ehrenreich and Janet McIntosh, in their 1997 article on 'The New Creationism' in the *Nation*, recount how a social psychologist called Phoebe Ellsworth was intimidated at an interdisciplinary seminar on emotions. Though bending over backwards to pre-empt criticism, at one point she unguardedly mentioned the word 'experiment'. Immediately, 'the hands shot up. Audience members pointed out that the experimental method is the brainchild of white Victorian males.' Carrying conciliation to what would have seemed to me almost superhuman lengths, Ellsworth agreed that white males had done their share of damage in the world but noted that, none the less, their efforts had led to the discovery of DNA. This earned the incredulous (and incredible) retort: 'You believe in DNA?' Fortunately, there are still many intelligent young women prepared to enter a scientific career, and I should like to pay tribute to their courage, in the face of uncouth bullying of this kind.

Of course a form of feminist influence in science is admirable and overdue. No well-meaning person could oppose campaigns to improve the status of women in scientific careers. It is truly appalling (as well as desperately sad) that Rosalind Franklin, whose X-ray diffraction photographs of DNA crystals were crucial to Watson and Crick's success, was not allowed in the common room of her own institution and was therefore debarred from contributing to, and learning from, what might have been crucial scientific shop talk. It also may be true that women typically can bring a point of view to scientific discussions which men typically do not. But 'typically' is not the same thing as 'universally', and the scientific truths that men and women eventually discover (albeit there may be statistical differences in the kinds of research that they are drawn to) will be accepted equally by reasonable people of both sexes, once they have been clearly established by members of either sex. And no, reason and logic are *not* masculine instruments of oppression. To suggest that they are is an insult to women, as Steven Pinker has said:

Among the claims of 'difference feminists' are that women do not engage in abstract linear reasoning, that they do not treat ideas with skepticism or evaluate them through rigorous debate, that they do not argue from general moral principles, and other insults.

How the Mind Works (1998)

The most ridiculous example of feminist bad science may be Sandra Harding's description of Newton's *Principia* as a 'rape manual'. What strikes me about this judgement is less its presumption than its parochial American chauvinism. How *dare* she elevate her narrowly contemporary North American politics over the unchanging laws of the universe and one of the greatest thinkers of all time (who happened, incidentally, to be male and rather unpleasant)? Paul Gross and Norman Levitt discuss this and similar examples in their admirable book *Higher Superstition* (1994), leaving the last word to the philosopher Margarita Levin:

... much of feminist scholarly writing consists of wildly extravagant praise of other feminists. A's 'brilliant analysis' supplements B's 'revolutionary breakthrough' and C's 'courageous undertaking.' More disconcerting is the penchant of many feminists to praise themselves most fulsomely. Harding ends her book on the following self-congratulatory note: 'When we began theorizing our experience ... we knew our task would be a difficult though exciting one. But I doubt that in our wildest dreams we ever imagined we would have to reinvent both science and theorizing itself to make sense of women's social experience.' This megalomania would be disturbing in a Newton or Darwin: in the present context it is merely embarrassing.

In the rest of this chapter I shall deal with various examples of bad poetic science drawn from my own field of evolutionary theory. The first, which not all would regard as bad science and which can be defended, is the vision of Herbert Spencer, Julian Huxley and others (including Teilhard de Chardin) of a general law of progressive evolution working at all levels in nature, not just the biological level. Modern biologists use the word evolution to mean a rather carefully defined process of systematic shifts in gene frequencies in populations, together with the resulting changes in what animals and plants actually look like as the generations go by. Herbert Spencer, who, to be fair, was the first to use the word evolution in a technical sense, wanted to regard biological evolution as only a special case. Evolution, for him, was a much more general process, with shared laws at all its levels. Other manifestations of the same general law of evolution were the development of the individual (the progression from fertilized egg through foetus to adult); the development of the cosmos, the stars and the planets from simpler beginnings; and progressive changes, over historical time, in social phenomena such as the arts, technology and language.

There are good things and bad about the poetry of general evolutionism. On balance I think it fosters confusion more than illumination, but there is certainly some of both. The analogy between embryonic development and species evolution was artfully exploited by that irascible genius J. B. S. Haldane to make a debating point. When a sceptic of evolution doubted that anything so complicated as a human could have come from single-celled beginnings, Haldane promptly observed that the sceptic himself had done that very thing and the whole process took only nine months. Haldane's rhetorical point is undiminished by the fact, which of course he knew perfectly well, that development is not the same thing as evolution. Development is change in the form of a single object, as clay deforms under a potter's hands. Evolution, as seen in fossils taken from successive strata, is more like a sequence of frames in a cinema film. One frame doesn't literally change into the next, but we experience an illusion of change if we project the frames in succession. With this distinction in place, we can quickly see that the cosmos does not evolve (it develops) but technology does evolve (early aeroplanes are not moulded into later ones but the history of aeroplanes, and of many other pieces of technology, falls well into the cinema frame analogy). Clothes fashions, too, evolve rather than develop. It is controversial whether the analogy between genetic evolution, on the one hand, and cultural or technical evolution, on the other, leads to illumination or the reverse, and I am not going to get into that argument now.

My remaining examples of bad poetry in evolutionary science come largely from a single author, the American paleontologist and essayist Stephen Jay Gould. I am anxious that such critical concentration upon one individual shall not be taken as personally rancorous. On the contrary, it is Gould's excellence as a writer that makes his errors, when they occur, so eminently worth rebutting.

In 1977 Gould wrote a chapter on 'eternal metaphors of paleontology' to introduce a multi-authored book on the evolutionary study of fossils. Beginning with Whitehead's preposterous, though much quoted, statement that all of philosophy is a footnote to Plato, Gould's thesis, in the words of the preacher of Ecclesiastes (whom he also quotes), is that there is nothing new under the sun: 'The thing that hath been, it is that which shall be; and that which is done is that which shall be done.' Current controversies in paleontology are just old controversies being recycled. They

preceded evolutionary thought and found no resolution within the Darwinian paradigm ... Basic ideas, like idealized geometric figures, are few in number. They are eternally available for consumption ...

Gould's eternally unresolved questions in paleontology are three in number: Does time have a directional arrow? Is the driving motor of evolution internal or external? Does evolution proceed gradually or in sudden jumps? Historically, he finds examples of paleontologists who have espoused all eight possible combinations of answers to these three questions, and he satisfies himself that they straddle the Darwinian revolution as though it never happened. But he manages this feat only by forcing analogies between schools of thought which, carefully examined, have no more in common than blood and wine, or helical orbits and helical DNA. All three of Gould's eternal metaphors are bad poetry, forced analogies that obscure rather than illuminate. And bad poetry in his hands is all the more damaging because Gould is a graceful writer.

The question whether evolution has a directional arrow is certainly one that can sensibly be asked, in various guises. But the bedfellows that the different guises bring together are so ill matched that they are not usefully united. Does bodily structure get progressively more complex as evolution goes on? This is a reasonable question. So is the question of whether the total diversity of species on the planet increases progressively as the ages go by. But they are utterly different questions and it is conspicuously unhelpful to invent a century-spanning school of 'progressivist' thought to unite them. Still less do either of them, in their modern form, have anything in common with the pre-Darwinian schools of 'vitalism' and 'finalism', which held that living things were progressively 'driven' from within, by some mystical life force, towards an equally mystical final goal. Gould forces unnatural connections among all these forms of progressivism, as a device to support his poetic historical thesis.

Much the same is true of the second eternal metaphor, and the question of whether the motor of change is in the external environment, or whether change arises from 'some independent and internal dynamic within organisms themselves'. A prominent modern disagreement is between those who believe that the main driving force of evolution is Darwinian natural selection and those who emphasize other forces such as random genetic drift. This important distinction is not conveyed, not even to the smallest extent, by the internalist/externalist dichotomy that Gould would force upon us in order to maintain his thesis that post-Darwinian argumentation is just a recycling of pre-Darwinian equivalents. Is natural selection externalist or internalist? It depends whether you are talking about adaptation to the external environment or co-adaptation of the parts to each other. I shall return to this distinction later in another context.

Bad poetry is even more evident in Gould's exposition of the third of his eternal metaphors, the one concerning gradual versus episodic evolution. Gould uses the word episodic to unite three kinds of sharp discontinuity in evolution. These are: first, catastrophes such as the mass extinction of the dinosaurs; second, macromutations or saltations; and third, punctuation in the sense of the theory of punctuated equilibrium proposed by Gould and his colleague Niles Eldredge in 1972. This last theory needs more explanation, and I'll come to it in a moment.

Catastrophic extinctions are straightforward to define. Exactly what causes them is controversial and probably different in different cases. For the moment, just notice that a worldwide catastrophe in which most species die is, to put it mildly, not the same thing as a macromutation. Mutations are random errors in gene copying and macromutations are mutations of large effect. A mutation of small effect, or micromutation, is a small error in gene copying, whose effect on its possessors might be too slight to notice easily, say a subtle lengthening of a leg bone, or a hint of reddening in a feather. A macromutation is a dramatic error, a change so large that, in extreme cases, its possessor would be classified in a different species from its parents. In my previous book, *Climbing Mount Improbable*, I reproduced a photograph from a newspaper of a toad with eyes in the roof of its mouth. If this photograph is genuine (a big if, in these days of Photoshop and other handy image-manipulation software), and if the error is genetic, the toad is a macromutant. If such a macromutant spawned a new species of toads with eyes in the roofs of their mouths, we should describe the abrupt evolutionary origin of the new species as a saltation or evolutionary jump. There have been biologists, such as the German/American geneticist Richard Goldschmidt, who believed that such saltatory steps were important in natural evolution. I am one of many who have cast doubt on the general idea, but that is not my purpose here. Here I make the much more basic point that such genetic leaps, even if they occur, have nothing in common with earth-shattering catastrophes such as the sudden extinction of the dinosaurs, except that both are sudden. The analogy is purely poetic, and it is bad poetry which doesn't lead to any further illumination. Recalling Medawar's words, the analogy marks the end, not the inauguration, of a train of thought. The ways of being a non-gradualist are so varied as to strip the category of all usefulness.

The same applies to the third category of non-gradualists: punctuationists in the sense of Eldredge and Gould's theory. The idea here is that a species comes into existence in a time which is short compared with the much longer period of 'stasis' during which it survives unchanged after its initial formation. In the extreme version of the theory, the species, once it has burst into existence, continues unchanged until either it goes extinct or it splits to form a new daughter species. It is when we ask what happens during the sudden bursts of species formation that the confusion, born of bad poetry, arises. There are two things that might happen. They are utterly different from each other, but Gould makes light of the difference because he is seduced by bad poetry. One is macromutation. The new species is founded by a freak individual, like the alleged toad with eyes in the roof of its mouth. The other thing that might happen – more plausibly, in my view, but I'm not talking about that now – is what we can call rapid gradualism. The new species comes into existence in a brief episode of rapid evolutionary change which, although gradual in the sense that parents don't spawn an instant new species in a single generation, is fast enough to look like an instant in the fossil record. The change is spread over many generations of small, step-by-step increments, but it looks like a sudden jump. This is either because the intermediates lived in a different place (say, on an outlying island) and/or because the intermediate stages passed too rapidly to fossilize – 10,000 years is too short to measure in many geological strata, yet it constitutes ample time for quite major evolutionary change to accumulate gradually in small steps.

There is all the difference in the world between rapid gradualism and macromutational saltation. They depend upon totally different mechanisms and they have radically different implications for Darwinian controversies. To lump them together simply because, like catastrophic extinctions, they all lead to discontinuities in the fossil record, is bad poetic science. Gould is aware of the difference between rapid gradualism and macromutation, but he treats the matter as though it were a minor detail, to be cleared up after we have taken on board the overarching question of whether evolution is episodic rather than gradual. One can see it as overarching only if one is intoxicated by bad poetry. It makes as little sense as my correspondent's question about the DNA double helix and whether it 'comes from' the earth's orbit. Once again, rapid gradualism no more resembles macromutation than a bleeding wizard resembles a shower of rain.

Even worse is to claim catastrophism under the same punctuationist umbrella. In pre-Darwinian times the existence of fossils became increasingly embarrassing for upholders of biblical creation. Some hoped to drown the problem in Noah's flood, but why did the strata seem to show dramatic replacements of whole faunas, each one different from its predecessor, and all of them largely free of our own, familiar creatures? The answer given by, among others, the nineteenth-century French anatomist Baron Cuvier, was catastrophism. Noah's flood was only the last in a series of cleansing disasters visited upon the earth by a supernatural power. Each catastrophe was followed by a new creation.

Apart from the supernatural intervention, this has something – a little – in common with our modern belief that mass extinctions such as those that ended the Permian and Cretaceous eras were followed by new flowerings of evolutionary diversity to match previous radiations. But to lump the catastrophists in with macromutationists and with modern punctuationists, just because all three can be represented as non-gradualist, is very bad poetry indeed.

After giving lectures in the United States, I have often been puzzled by a certain pattern of questioning from the audience. The questioner calls my attention to the phenomenon of mass extinction, say, the catastrophic end of the dinosaurs and their succession by the mammals. This interests me greatly and I warm to what promises to be a stimulating question. Then I realize that the tone of the question is unmistakably *challenging*. It is almost as though the questioner expects me to be surprised, or discomfited, by the fact that evolution is periodically interrupted by catastrophic mass extinctions. I was baffled by this until the truth suddenly hit me. Of course! The questioner, like many people in North America, has learned his evolution from Gould, and I have been billed as one of those 'ultra-Darwinian' *gradualists*! Doesn't the comet that killed the dinosaurs also blow my gradualistic view of evolution out of the water? No, of course it doesn't. There is not the smallest connection. I am a gradualist in the sense that I don't think macromutations have played an important role in evolution. More determinedly, I am a gradualist when it comes to explaining the evolution of complex adaptations like eyes (so is any sane person, including Gould). But what on earth have such matters got to do with mass extinctions? Nothing at all. Unless, that is, your mind has been filled up with bad poetry. For the record, I believe, and have believed for the whole of my career, that mass extinctions exert a profound and dramatic influence on the subsequent course of evolutionary history. How could they not? But mass extinctions are not a part of the Darwinian process, except in so far as they clear the decks for new Darwinian beginnings.

There is irony lurking here. Among the facts about extinction that Gould is fond of emphasizing is its capriciousness. He calls it contingency. When mass extinction strikes, major groups of animals die wholesale. In the Cretaceous extinction, the once mighty group of dinosaurs (with the notable exception of birds) was completely wiped out. The choice of major group for victim is either random or, if non-random, it is not the same non-randomness as we see in conventional natural selection. The normal adaptations to survival do not avail against comets. Grotesquely, this fact is sometimes trotted out as if it were a debating point against neo-Darwinism. But neo-Darwinian natural selection is selection *within* species, not between species. To be sure, natural selection involves death, and mass extinction involves death, but any further resemblance between the two is purely poetic. Ironically, Gould is one of the few Darwinians who still think of natural selection as working at levels higher than the individual organism. It would never occur to the rest of us even to *ask* whether mass extinctions are selective events. We might see extinction as opening up new opportunities for adaptation, by lower-level natural selection choosing between individuals separately within each species that has survived the catastrophe. As a further irony, it is the poet Auden who came nearer to getting it right:

But catastrophes only encouraged experiment.
As a rule, it was the fittest who perished, the mis-fits,
forced by failure to emigrate to unsettled niches, who
altered their structure and prospered.

'Unpredictable but Providential (for Loren Eiseley)'

I take one further extended example of bad poetic science from paleontology, and once again Stephen Jay Gould is responsible for its popularity even if he has not clearly expressed it himself in its extreme form. Many readers of his elegantly written book *Wonderful Life* (1989) have been captivated by the idea that there was something special and unique about the whole business of evolution in the Cambrian era, when fossils of most of the great animal groups first appeared, rather over 500 million years ago. It is not just that the animals of the Cambrian were peculiar. Of course they were. The animals of every era have their peculiarities and the Cambrian ones were arguably more peculiar than most. No, the suggestion is that the whole process of evolution in the Cambrian was odd.

The standard neo-Darwinian view of the evolution of diversity is that a species splits into two when two populations become sufficiently unalike that they can no longer interbreed. Often the populations begin diverging when they chance to be geographically separated. The separation means that they no longer mix their genes sexually and this permits them to evolve in different directions. The divergent evolution might be driven by natural selection (which is likely to push in different directions because of different conditions in the two geographical areas). Or it might consist of random evolutionary drift (since the two populations are not genetically held together by sexual mixing, there is nothing to stop them drifting apart). In either case, when they have evolved sufficiently far apart that they could no longer interbreed even if they were geographically united again, they are defined as belonging to separate species.

Subsequently, the lack of interbreeding permits further evolutionary divergence. What had been distinct species within one genus become, in the fullness of time, distinct genera within one family. Later, families will be found to have diverged to the point where taxonomists (specialists in classification) prefer to call them orders, then classes, then phyla. Phylum (plural phyla) is the classificatory name by which we distinguish really fundamentally different animals like molluscs, nematode worms, echinoderms and chordates (chordates are mostly vertebrates plus a few odds and ends). Ancestors of two different phyla, say vertebrates and molluscs, which we see as built upon utterly different 'fundamental body plans' were once just two species within a genus. Before that, they were two geographically separated populations within one ancestral species. The implication of this widely accepted view is that, as you go back and back in geological time, the gap between any pair of animal groups becomes smaller and smaller. The further back in time you go, the closer you approach the uniting of these different kinds of animals in their single common ancestor species. Our ancestors and mollusc ancestors were once very alike. Later they were not quite so alike. Later again they had diverged further, and so on until eventually they became so different that we should call them two phyla. This general story can scarcely be doubted by any reasonable person who thinks it through, though we do not have to be committed to the view that it occurs at a uniform rate with time. It could have happened in rapid bursts.

The dramatic phrase 'Cambrian explosion' is used in two senses. It can refer to the factual observation that before the Cambrian era, just over half a billion years ago, there are few fossils. Most of the great animal phyla appear as fossils for the first time in Cambrian rocks, and this looks like a great explosion of new animals. The second meaning is the theory that the phyla actually branched off from each other during the Cambrian, even during as little as 10 million years within the Cambrian. This second idea, which I shall call the branch point explosion hypothesis, is controversial. It is compatible – just – with what I am calling the standard neo-Darwinian model of species divergence. We've already agreed that, as we trace any pair of modern phyla back in time, we eventually converge upon a common ancestor. My hunch is that, for different pairs of phyla, we'll hit the common ancestor in different geological eras: say, the common ancestor of vertebrates and molluscs at 800 million years ago, the common ancestor of vertebrates and echinoderms at 600 million years, and so on. But I could be wrong, and we can easily accommodate the branch point explosion hypothesis by saying that, for some reason (which is interesting enough to need investigating), most of our backward tracings happen to hit their respective common ancestors during the same relatively short geological period, say, between 540 million and 530 million years ago. This would have to mean that, at least near the beginning of that 10-million-year period, the ancestors of the modern phyla were nowhere near as different from each other as they are today. They were, after all, diverging from common ancestors at the time and were originally members of the same species.

The extreme Gouldian view – certainly the view inspired by his rhetoric, though it is hard to tell from his own words whether he literally holds it himself – is radically different from and utterly incompatible with the standard neo-Darwinian model. It also, as I shall show, has implications which, once they are spelled out, anybody can see are absurd. It is very clearly expressed – betrayed might be a better word – in asides in Stuart Kauffman's *At Home in the Universe* (1995):

One might imagine that the first multicellular creatures would all be very similar, only later diversifying, from the bottom up, into different genera, families, orders, classes, and so on. That, indeed, would be the expectation of the strictest conventional Darwinist. Darwin, profoundly influenced by the emerging view of geologic gradualism, proposed that all evolution occurred by the very gradual accumulation of useful variations. Thus the earliest multicellular creatures themselves ought to have diverged gradually from one another.

So far, this is a fine summary of the orthodox neo-Darwinian view. Now, in a bizarre passage, Kauffman goes on:

But this appears to be false. One of the wonderful and puzzling features of the Cambrian explosion is that the chart was filled in from the top down. Nature suddenly sprang forth with many wildly different body plans – the phyla – elaborating on these basic designs to form the classes, orders, families, and genera . . . In his book about the Cambrian explosion, Wonderful Life: The Burgess Shale and the Nature of History, *Stephen Jay Gould remarks on this top-down quality of the Cambrian with wonder.*

As well he might! You only have to think for one moment about what 'top down' filling in would have to mean for the animals on the ground and you immediately see how preposterous it is. 'Body plans' like the mollusc body plan, or the echinoderm body plan, are not ideal essences hanging in the sky, waiting, like designer dresses, to be adopted by real animals. Real animals is all there ever was: living, breathing, walking, eating, excreting, fighting, copulating real animals, who had to survive and who can't have been dramatically different from their real parents and grandparents. For a new body plan – a new phylum – to spring into existence, what actually has to happen on the ground is that a child is born which suddenly, out of the blue, is as different from its parents as a snail is from an earthworm. No zoologist who thinks through the implications, not even the most ardent saltationist, has ever supported any such notion. Ardent saltationists have been content to postulate the sudden bursting into existence of new *species*, and even that relatively modest idea has been highly controversial. When you spell out the Gouldian rhetoric into real-life practicalities, it stands revealed as the purest of bad poetic science.

Kauffman is even more explicit in a later chapter. In discussing some of his ingenious mathematical models of evolution on 'rugged fitness landscapes', Kauffman notes a pattern that he thinks

sounds a lot like the Cambrian explosion. Early on in the branching process, we find a variety of long-jump mutations that differ from the stem and from one another quite dramatically. These species have sufficient morphological differences to be categorized as founders of distinct phyla. These founders also branch, but do so via slightly closer long-jump variants, yielding branches from each founder of a phylum to dissimilar daughter species, the founders of classes. As the process continues, fitter variants are found in progressively more nearby neighborhoods, so founders of orders, families, and genera emerge in succession.

Kauffman's earlier, more technical book, *The Origins of Order* (1993), says something similar about life in the Cambrian:

Not only did a very large number of novel body forms originate rapidly, but the Cambrian explosion exhibited another novelty: Species which founded taxa appear to have built up the higher taxa from the top down. That is, exemplars of major phyla were present first, followed by progressive filling in at class, order, and lower taxonomic levels . . .

Now, one way of reading this is inoffensive to the point of obviousness. On our 'converging backwards' model it would have to be true that species splittings that are *eventually going to become* phylum divides would normally precede those that are destined to become divides between orders and lower taxonomic levels. But Kauffman clearly doesn't think he is saying something ordinary and obvious. This is apparent from his statement that 'the Cambrian explosion exhibited another novelty', and from his phrase 'long-jump mutations'. He thinks he is attributing to the Cambrian something revolutionary. He really does seem genuinely to intend the alternative reading, in which 'long-jump mutations' give rise, on the instant, to brand new phyla.

I hasten to emphasize that these passages of Kauffman's are embedded in a pair of books that are for the most part interesting, creative and uninfluenced by Gould. The same is true of Richard Leakey and Roger Lewin's *The Sixth Extinction* (1996), another recent book, admirable in most of its chapters, but sadly marred by one, 'The Mainspring of Evolution', which is explicitly and avowedly influenced by Gould. Here are a couple of relevant passages:

It was as if the facility for making evolutionary leaps that produced major functional novelties – the basis of new phyla – had somehow been lost when the Cambrian period came to an end. It was as if the mainspring of evolution had lost some of its power.

Hence, evolution in Cambrian organisms could take bigger leaps, including phylum-level leaps, while later on it would be more constrained, making only modest jumps, up to the class level.

As I have written before, it is as though a gardener looked at an old oak tree and remarked, wonderingly: 'Isn't it strange that no major new boughs have appeared on this tree for many years. These days, all the new growth appears to be at the twig level!' Just think once again what a 'phylum-level leap' or even a 'modest' (*modest?*) class level leap would have to mean. Animals of different phyla, remember, are animals with different fundamental body plans, like molluscs and vertebrates. Or like starfish and insects. A long-jump, phylum level mutation would have to mean that a couple of parents belonging to one phylum mated and gave birth to a child belonging to a different phylum. The difference between parent and offspring would have to be on the same scale as the difference between a snail and a lobster, or a starfish and a codfish. A class level leap would be equivalent to a pair of birds giving birth to a mammal. Picture the parents gazing wonderingly into the nest at what they have produced, and the full comedy of the notion becomes apparent.

My assurance in ridiculing these ideas is not based simply upon knowledge of the facts of modern animals. Obviously if it were just that, one could retort that things were different in the Cambrian. No, the argument against Kauffman's long jumps, or Leakey and Lewin's phylum level leaps, is a theoretical one, and an extremely strong one. It is this. Even if mutations on this gigantic scale occurred, the products would not have survived. This is fundamentally because, as I have said before, however many ways there may be of being alive, there are almost infinitely more ways of being dead. A small mutation, representing a minor step away from a parent which has proved its ability to survive by virtue of being a parent, has a good chance of surviving for the same reason, and it may even be an improvement. A gigantic, phylum level mutation is a leap into the wild blue yonder. I said that the long-jump mutation we are talking about would be of the same *magnitude* as a mutation from a mollusc to an insect. But it would never, of course, have *been* a jump from a mollusc to an insect. An insect is a highly tuned piece of survival machinery. If a mollusc parent gave birth to a new phylum, the leap would have been a random leap, like any other mutation. And the chance that a random leap of that magnitude would produce an insect, or *anything with the faintest chance of surviving*, is small enough to be discounted totally. The chance of its being viable is impossibly small, no matter how empty the ecosystem, how wide open the niches. A phylum level leap would be a mess.

I do not believe the authors I am quoting really believe what their printed words undoubtedly appear to be saying. I think they were simply intoxicated by Gould's rhetoric and didn't think it through. The whole point of quoting them in this chapter is to illustrate the power to mislead that a skilled poet can unwittingly exert, especially if he has first misled himself. And the poetry of the Cambrian as a blissful dawn of innovation is undoubtedly beguiling. Kauffman gets completely carried away by it:

Soon after multicelled forms were invented, a grand burst of evolutionary novelty thrust itself outward. One almost gets the sense of multicellular life gleefully trying out all its possible ramifications, in a kind of wild dance of heedless exploration.

At Home in the Universe (1995)

Yes. One does get *exactly* that sense. But one gets it from Gould's rhetoric, not from the facts of Cambrian fossils nor from sober reasoning about evolutionary principles.

If scientists of the calibre of Kauffman, Leakey and Lewin can be seduced by bad poetic science, what chance has the non-specialist? Daniel Dennett has told me of a conversation with a philosopher colleague who had read *Wonderful Life* as arguing that the Cambrian phyla did not have a common ancestor – that they had sprung up as independent origins of life! When Dennett assured him that this was not Gould's intention, his colleague's response was, 'Well then, what is all the fuss about?'

Excellence in writing is a double-edged sword, as the distinguished evolutionary scientist John Maynard Smith has noted, in the *New York Review of Books*, November 1995:

Gould occupies a rather curious position, particularly on his side of the Atlantic. Because of the excellence of his essays, he has come to be seen by non-biologists as the preeminent evolutionary theorist. In contrast, the evolutionary biologists with whom I have discussed his work tend to see him as a man whose ideas are so confused as to be hardly worth bothering with, but as one who should not be publicly criticized because he is at least on our side against the creationists. All this would not matter, were it not that he is giving non-biologists a largely false picture of the state of evolutionary theory.

Maynard Smith was reviewing Dennett's book *Darwin's Dangerous Idea* (1995), which contains a devastating and, one might hope, terminal critique of Gould's influence on evolutionary thinking.

What really happened in the Cambrian? Simon Conway Morris of Cambridge University is, as Gould fulsomely acknowledges, one of the three leading modern investigators of the Burgess Shale, the Cambrian fossil bed which is the subject of *Wonderful Life.* Conway Morris has recently published his own fascinating book on the subject, *The Crucible of Creation* (1998), which is critical of almost every aspect of Gould's view. Like Conway Morris, I don't think there's any good reason to think that the process of evolution was different in the Cambrian from the way it is today. But there is no doubt that a large number of major animal groups are seen in the fossil record for the first time in the Cambrian. The obvious hypothesis has occurred to many people. Perhaps several groups of animals evolved hard, fossilizable skeletons around the same time and perhaps for the same reason. One possibility is an evolutionary arms race between predators and prey, but there are other ideas like a dramatic change in the chemistry of the atmosphere. Conway Morris finds no support at all for the poetic idea of an exuberant and extravagant flowering of life in a wild dance of Cambrian diversity and disparity, subsequently pruned to today's more limited repertoire of animal types. If anything, the reverse seems to be true, as most evolutionists would expect.

Where does that leave the question of the timing of the branch points of the major phyla? Recall that this is a separate question from the undoubted Cambrian explosion of fossil availability. The controversial matter is whether the branch points in the divergence of all the major phyla are concentrated in the Cambrian – the branch point explosion hypothesis. I said standard neo-Darwinism was compatible with this hypothesis. But I still don't think it is at all likely.

One possible way to tackle the question is by looking at molecular clocks. 'Molecular clock' refers to the observation that certain biological molecules change at a rather fixed rate over the millions of years. If you accept this, you can take blood from any two modern animals and calculate how long ago their common ancestor lived. Some recent molecular clock studies have pushed the branch points of various pairs of phyla deep into the Precambrian era. If these studies are right, the whole rhetoric of an evolutionary explosion becomes superfluous. But there is controversy over the interpretation of molecular clock results so far back in deep time, and we should wait for more evidence.

Meanwhile, there is a logical argument which I can assert with more confidence. The only evidence in favour of the branch point explosion hypothesis is negative: there aren't any fossils of many of the phyla before the Cambrian. But those fossil animals that have no fossil ancestors must have had ancestors of some kind. They can't have sprung from nothing. *Therefore there must have been ancestors that didn't fossilize*, absence of fossils does not mean absence of animals. The only question that remains is whether the missing ancestors going back to the branch points, *who must have existed*, were all compressed into the Cambrian, or whether they were strung out through the previous hundreds of millions of years. Since the only reason to suppose that they were compressed into the Cambrian is the absence of fossils, and since we have just proved logically the irrelevance of that absence, I conclude that there is no good reason at all to favour the branch point explosion hypothesis. Doubtless it has great poetic appeal.

9

THE SELFISH COOPERATOR

Wonder . . . and not any expectation of advantage from its discoveries, is the first principle which prompts mankind to the study of Philosophy, of that science which pretends to lay open the concealed connections that unite the various appearances of nature.

ADAM SMITH,[6] The History of Astronomy' (1795)

The medieval bestiaries continued an earlier tradition of hijacking nature as a source of moral tales. In its modern form, in the development of evolutionary ideas, the same tradition underlies one of the most egregious forms of bad poetic science. I refer to the illusion that there is a simple opposition between nasty and nice, social and antisocial, selfish and altruistic, tough and gentle; that these pairs of binary opposites all correspond to the other pairs, and that the history of evolutionary controversy about society is described by a pendulum swinging back and forth along a continuum between these opposites. I am not denying that there are interesting issues to be discussed hereabouts. What I am criticizing is the 'poetic' idea that there is a single continuum and that worthwhile arguments are to be had between vantage points along its length. To invoke the rainmakers yet again, there is no more connection between a selfish gene and a selfish human than there is between a rock and a rain cloud.

To explain the poetic continuum I am criticizing, I might as well borrow a line from a real poet, Tennyson's 'Nature, red in tooth and claw', from *In Memoriam* (1850), widely assumed to be inspired by *On the Origin of Species* but actually published nine years earlier. At one end of the poetic continuum are supposed to stand Thomas Hobbes, Adam Smith, Charles Darwin, T. H. Huxley and all those, such as the distinguished American evolutionist George C. Williams and today's advocates of 'the selfish gene', who emphasize that nature really is red in tooth and claw. At the other end of the continuum are Prince Peter Kropotkin the Russian anarchist and author of *Mutual Aid* (1902), the gullible but immensely influential American anthropologist Margaret Mead,* and today a spate of authors reacting indignantly to the idea that nature is genetically selfish, of whom Frans de Waal, author of *Good Natured* (1996) is representative.

De Waal, a chimpanzee expert who understandably loves his animals, is distressed at what he mistakenly sees as a neo-Darwinian tendency to emphasize the 'nastiness of our apish past'. Some of those who share his romantic fancy have recently become fond of the pygmy chimpanzee or bonobo as a yet more benign role model. Where common chimpanzees often resort to violence, and even cannibalism, bonobos say it with sex. They seem to copulate in all possible combinations at every conceivable opportunity. Where we might shake hands, they copulate. Make love not war is their watchword. Margaret Mead would have warmed to them. But the very idea of taking animals to be role models, as in the bestiaries, is a piece of bad poetic science. Animals are not there to be role models, they are there to survive and reproduce.

Moralistic devotees of the bonobo are apt to compound this error

* I should explain that Margaret Mead is 'gullible but influential' because a large section of American academic culture enthusiastically adopted her rose-tinted environmentalist theory of human nature which, it later transpired, she had built on a somewhat insecure foundation: systematic misinformation fed her as a joke by two mischievous Samoan girls, during her brief period of fieldwork in their island. She didn't stay in Samoa long enough to learn the language well, unlike her professional nemesis, the Australian anthropologist Derek Freeman, who uncovered the whole story years later in the course of a more detailed study of Samoan life.

with an outright evolutionary falsehood. Probably because of their powerful 'feelgood factor', bonobos are often claimed as more closely related to us than common chimpanzees are. But this cannot be, so long as we accept, as everybody does, that bonobos and common chimpanzees are more closely related to each other than either is to humans. You need no more than that simple and uncontroversial premise to conclude that bonobos and common chimpanzees are exactly *equally* closely related to us. They are connected to us via the common ancestor that they share and we don't. Certainly we may resemble one of the two species more than the other in some respects (and very probably resemble the other in other respects), but such comparative judgements absolutely cannot be reflections of differential evolutionary closeness.

De Waal's book is full of anecdotal demonstrations (which should surprise nobody) that animals are sometimes kind to each other, cooperate for mutual good, care for each other's welfare, console each other in distress, share food and do other heartwarmingly good things. The position I have always adopted is that much of animal nature is indeed altruistic, cooperative and even attended by benevolent subjective emotions, but that this follows from, rather than contradicts selfishness at the genetic level. Animals are sometimes nice and sometimes nasty, since either can suit the self-interest of genes at different times. That is precisely the reason for speaking of 'the selfish gene' rather than, say, 'the selfish chimpanzee'. The opposition that de Waal and others have erected, between biologists who believe human and animal nature is fundamentally selfish, and those who believe it is fundamentally 'good-natured', is a false opposition – bad poetry.

It is now widely understood that altruism at the level of the individual organism can be a means by which the underlying genes maximize their self-interest. However, I don't want to dwell on what I have expounded in earlier books such as *The Selfish Gene.* What I would now re-emphasize from that book – it has been overlooked by critics who appear to have read it by title only – is the important sense in which genes, though in one way purely selfish, at the same time enter into cooperative cartels with each other. This is poetic science, if you like, but I hope to show that it is good poetic science which aids understanding rather than impedes it. I shall do the same with other examples in the remaining chapters.

The key insight of Darwinism can be expressed in genetic terms. The genes that exist in many copies in the population are the ones that are good at making copies, which also means good at surviving. Surviving where? Surviving in individual bodies in ancestral environments. That means surviving in the environment typical of the species: in a desert for camels, up trees for monkeys, in the deep sea for giant squids, and so on. The reason individual bodies are so good at surviving in their environments is mainly that they have been built by genes that have survived in the same environment for many generations, in the form of copies.

But never mind deserts and ice floes, seas and forests; they are only part of the story. A far more prominent aspect of the ancestral environment in which genes have survived is the other genes with which they have had to share a succession of individual bodies. The genes that survive in camels will, to be sure, include some that are particularly good at surviving in deserts, and they may even be shared with desert rats and desert foxes. But more importantly, successful genes will be those that are good at surviving in an environment consisting of the other genes that are typically found in the species. So, the genes of a species become selected to be good at cooperating with each other. Genetic cooperation, which is good scientific poetry whereas universal cooperation is not, will be the subject of this chapter.

The following fact is often misunderstood. It is not the genes of any given individual that cooperate particularly well together. They have never been together before in that combination, for every genome in a sexually reproducing species is unique (with the usual exception of identical twins). It is the genes of a species at large that cooperate, because they have met before, often, and in the intimately shared environment of the cell, though always in different combinations. What they cooperate at is the business of making individuals of the same general type as the present one. There is no particular reason to expect the genes of any particular individual to be especially good at cooperating with one another when compared with any other genes of the same species. It is largely a matter of accident which particular companions the lottery of sexual reproduction has drawn for them from the gene pool of the species. Individuals with unfavourable combinations of genes tend to die. Individuals with favourable combinations tend to pass those genes on to the future. But it is not the favourable combinations themselves that are passed on in the long term. Sexual reshuffling sees to that. Instead, what are passed on are the genes that tend to be good at forming favourable combinations with the other genes that the species gene pool has to offer. Over the generations, whatever else the surviving genes may be good at, they will be good at working together with other genes of the species.

For all we know, particular camel genes might be good at cooperating with particular cheetah genes. But they are never called upon to do so. Presumably mammal genes are better at cooperating with other mammal genes than with bird genes. But the speculation must remain hypothetical, because one of the characteristics of life on our planet is that, genetic engineering aside, genes are mixed only within species. We can test watered-down versions of such speculations by looking at hybrids. Hybrids between different species, when they exist at all, often survive less well, or are less fertile, than pure-bred individuals. At least part of the reason for this is incompatibilities between their genes. Species A genes that work well against a genetic background or 'climate' of other Species A genes do not work when transplanted into Species B, and vice versa. Similar effects are sometimes seen when varieties or races within one species hybridize.

I first understood this while listening to lectures by the late E. B. Ford, legendary Oxford aesthete and eccentric founder of the now neglected school of Ecological Geneticists. Most of Ford's research was on wild populations of butterflies and moths. Among these was the Lesser Yellow Underwing moth, *Triphaena comes*. This moth is normally yellowish brown, but there is a variant called *curtisii* which is blackish. *Curtisii* is not found in England at all; however, in Scotland and the Isles *curtisii* coexists with the normal *comes*. The *curtisii* dark colour pattern is nearly completely dominant to the normal *comes* pattern. 'Dominant to' is a technical term, which is why I can't just say 'dominates'. It means that hybrids between the two look like *curtisii* even though they bear the genes of both. Ford caught specimens from Barra in the Outer Hebrides, west of Scotland, and from one of the Orkney islands, north of Scotland, as well as from the Scottish mainland itself. Each of the two island forms looks exactly like its opposite number at the other island site, and the dark *curtisii* gene is dominant on the two islands, as well as on the mainland. Other evidence shows that the *curtisii* gene is the very same gene in all localities. In view of this you'd expect that, when you cross-bred specimens from different islands, the normal dominance pattern would hold up. But it doesn't, and this is the point of the story. Ford caught individuals from Barra and mated them with individuals from Orkney. And the dominance of *curtisii* completely disappeared. A full range of intermediates turned up in the hybrid families, just as if there was no dominance.

What seems to be going on is this. The *curtisii* gene does not, in itself, encode the formula for the coloured pigment by which we distinguish the moths, nor is dominance ever a property of a gene on its own. Instead, like any other gene, the *curtisii* gene should be thought of as having its effects only in the context of a suite of other genes, some of which it 'switches on'. This suite of other genes is part of what I mean by 'genetic background' or 'genetic climate'. In theory, any gene could therefore exert radically different effects on different islands, in the presence of different suites of other genes. In the case of Ford's Yellow Underwings, things are a little more complicated, and very illuminating. The *curtisii* gene is a 'switch gene' which has what looks like the same effect on both Barra and Orkney, but it achieves it by switching on different suites of genes on the different islands. We notice this only when the two populations are cross-bred. The *curtisii* switch gene finds itself in a genetic climate which is neither one thing nor the other. It is a mixture of Barra genes and Orkney genes, and the colour pattern that either suite could produce, on its own, breaks down.

What is interesting about this is that either the Barra mixture or the Orkney mixture can put together the colour pattern. There is more than one way of achieving the same result. Both of them involve cooperating suites of genes, but they are two different suites and the members of each suite don't cooperate well with the other. I take this to be a model for what often goes on among working genes within any gene pool. In *The Selfish Gene,* I used a rowing analogy. A crew of eight oarsmen needs to be well coordinated. Eight men who have trained together can expect to work well together. But if you mix four men from one crew with four from another equally good crew, they don't gel: their rowing falls apart. This is analogous to the mixing of two suites of genes which worked well when each was with its previous companions, but whose coordination breaks down when each is pushed into the foreign genetic climate provided by the other.

Now at this point many biologists get carried away and say that natural selection must work at the level of the whole crew as a unit, the whole suite of genes, or the whole individual organism. They are right that the individual organism is a very important unit in the hierarchy of life. And it really does display unitary qualities. (This is less true of plants than of animals, who tend to have a fixed set of parts, all neatly parcelled inside a skin with a discrete, unitary shape. Individual plants are often harder to delimit as they straggle and vegetatively propagate themselves through meadows and undergrowth.) But, however unitary and discrete an individual wolf or buffalo, say, may be, the package is temporary and it is unique. Successful buffaloes don't duplicate themselves around the world in the form of multiple copies, they duplicate their genes. The true unit of natural selection has to be a unit of which you can say it has a frequency. It has a frequency which goes up when its type is successful, down when it fails. This is exactly what you can say of genes in gene pools. But you can't say it of individual buffaloes. Successful buffaloes don't become more frequent. Each buffalo is unique. It has a frequency of one. You can define a buffalo as successful if its genes increase in frequency in future populations.

Field Marshal Montgomery, never the humblest of men, was once heard to remark, 'Now God said (and I agree with Him) . . .' I feel a bit like that when I read of God's covenant with Abraham. He didn't promise Abraham eternal life as an individual (though Abraham was only 99 at the time, a spring chicken by Genesis standards). But he did promise him something else.

And I will make my covenant between me and thee, and will multiply thee exceedingly . . . and thou shalt be a father of many nations . . . And I will make thee exceeding fruitful, and I will make nations of thee, and kings shall come out of thee. Genesis 17

Abraham was left in no doubt that the future lay with his seed, not his individuality. God knew his Darwinism.

To resume, the point I am making is that genes, for all that they are the separate units naturally selected in the Darwinian process, are highly cooperative. Selection favours or disfavours single genes for their capacity to survive in their environment, but the most important part of that environment is the genetic climate furnished by other genes. The consequence is that cooperating suites of genes come together in gene pools. Individual bodies are as unitary and coherent as they are, not because natural selection chooses them as units, but because they are built by genes that have been selected to cooperate with other members of the gene pool. They cooperate specifically in the enterprise of building individual bodies. But it is an anarchistic, 'each gene for itself' kind of cooperation.

The cooperation, indeed, breaks down whenever the chance arises, as in so-called 'segregation distorter' genes. There is a gene in mice known as the *t* gene. In double dose *t* causes sterility or death, and there must be strong natural selection against it. But in single dose in males, it has a very odd effect. Normally, each copy of a gene should find itself in 50 per cent of the sperms made by a male. I have brown eyes like my mother, but my father has blue, so I know that I carry one copy of the gene for blue eyes and 50 per cent of my sperms carry the blue-eyed gene. In male mice, *t* doesn't behave in this orderly way. More than 90 per cent of an affected male's sperms contain *t*. Distorting sperm production is what the *t* gene does. It is its equivalent of making brown eyes or curly hair. And you can see that, in spite of lethality in double dose, once *t* arises in a population of mice, it will tend to spread because of its huge success in getting itself into sperms. It has been suggested that outbreaks of *t* arise in wild populations of mice, spreading like a sort of population cancer and eventually driving the local population extinct. *t* is an illustration of what can happen when cooperation among genes breaks down. 'The exception that proves the rule' is often a rather silly expression, but this is a rare occasion when it is appropriate.

To repeat, the main suites of cooperating genes are the whole gene pools of species. Cheetah genes cooperate with cheetah genes but not with camel genes, and vice versa. This is not because cheetah genes, even in the most poetic sense, see any virtue in the preservation of the cheetah species. They are not working to save the cheetah from extinction like some molecular World Wildlife Fund. They are simply surviving in their environment, and their environment largely consists of other genes from the cheetah gene pool. Therefore, abilities to cooperate with other cheetah genes (but not with camel genes or codfish genes) are among the main qualities favoured in the struggle between rival cheetah genes. Just as, in arctic climates, genes for withstanding the cold come to predominate, so, in cheetah gene pools, do genes that are equipped to flourish in the climate of other cheetah genes. As far as each gene is concerned, the other genes in its gene pool are just another aspect of the weather.

The level at which the genes constitute 'weather' for each other is mostly buried in cellular chemistry. Genes code the production of enzymes, protein molecules which work as machine tools churning out one particular component in a chemical production line. There are alternative chemical pathways to the same end, which means alternative production lines. It may not matter greatly which of two production lines is adopted, so long as the cell doesn't attempt both at once. Either of the two production lines might be equally good, but intermediate products yielded by production line A can't be used by production line B, and vice versa. Once again, it is tempting to say that the entire production line is naturally selected, as a unit. This is wrong. What is naturally selected is each individual gene, against the background or climate provided by all the other genes. If the population happens to be dominated by genes for all but one of the steps in production line A, this constitutes a chemical climate in which the gene for the missing A step is favoured. Conversely, a pre-existing climate of B genes favours B genes over A genes. We aren't talking about which is 'better', as though there were some kind of contest between production line A and production line B. What we are saying is that either of the two is fine, but a mixture is unstable. The population has two alternative stable climates of mutually cooperating genes and natural selection will tend to steer the population towards whichever of the two stable states it is already closest to.

But we don't have to talk of biochemistry. We can use the metaphor of genetic climate at the level of organs and behaviour. A cheetah is a beautifully integrated killing machine, equipped with long, muscled legs and a sinuously sprung backbone for outrunning prey, powerful jaws and dagger teeth for stabbing them, forward-focused eyes for aiming at them, short guts with appropriate enzymes for digesting them, a brain pre-loaded with carnivorous behaviour software, and collections of other features that make it a typical hunter. On the other side of the arms race, antelopes are equivalently well equipped to eat plants and avoid being caught by predators. Long guts, complicated by blind alleys stuffed with cellulose-digesting bacteria, go with flat grinding teeth, go with brains pre-programmed to alarm and rapid escape, go with exquisitely camouflaged dappling of the pelt. These are two alternative ways of making a living. Neither is obviously better than the other, but either is better than an uneasy compromise: carnivorous guts combined with herbivorous teeth, say, or carnivorous pursuit instincts combined with herbivorous digestive enzymes.

Yet again, it is tempting to speak of the 'whole cheetah' or the 'whole antelope' as being selected 'as a unit'. Tempting, but superficial. Also lazy. It requires some extra thinking work to see what is really going on. Genes that programme the development of carnivorous guts flourish in a genetic climate that is already dominated by genes programming carnivorous brains. And vice versa. Genes that programme defensive camouflage flourish in a genetic climate that is already dominated by genes programming herbivorous teeth. And vice versa. There are lots and lots of ways of making a living. To mention only a few mammal examples, there is the cheetah way, the impala way, the mole way, the baboon way, the koala way. There is no need to say that one way is better than any other. All of them work. What is bad is to be caught with half your adaptations aimed at one way of life, half aimed at another.

This kind of argument is best expressed at the level of the separate genes. At each genetic locus, the gene most likely to be favoured is the one that is compatible with the genetic climate afforded by the others, the one that survives in that climate through repeated generations. Since this applies to each one of the genes that constitute the climate – since every gene is potentially part of the climate of every other – the result is that a species gene pool tends to coalesce into a gang of mutually compatible partners. Sorry to go on about this, but some of my respected colleagues refuse to get the point, obstinately insisting that the 'individual' is the 'true' unit of natural selection!

More widely, the environment in which a gene has to survive includes the other species with which it comes into contact. The DNA of any one species doesn't literally come into direct contact with the DNA molecules of its predators, competitors or mutualistic partners. 'Climate' has to be understood less intimately than when the arena of gene cooperation is the interior of cells, as it is for genes within one species. In the larger arena, it is the *consequences* of genes in other species – their 'phenotypic effects' – that constitute an important part of the environment in which the natural selection of genes within neighbouring species goes on. A rainforest is a special kind of environment, fashioned and defined by the plants and animals that live in it. Every one of the species in a tropical rainforest consists of a gene pool, isolated from all other gene pools as far as sexual mixing is concerned, but in contact with their bodily effects.

Within each of those separate gene pools, natural selection favours those genes that cooperate within their own gene pool, as we have seen. But it also favours those genes that are good at surviving alongside the consequences of the other gene pools in the rainforest – the trees, vines, monkeys, dung beetles, woodlice and soil bacteria. In the long run this may make the whole forest look like a single harmonious whole, with each unit pulling for the benefit of all, every tree and every soil mite, even every predator and every parasite, playing its part in one big, happy family. Once again, this is a tempting way of looking at it. Once again, it is lazy – bad poetic science. A much truer vision, still poetic science but (it is the purpose of this chapter to persuade you) *good* poetic science, sees the forest as an anarchistic federation of selfish genes, each selected as being good at surviving within its own gene pool against the background of the environment provided by all the others.

Yes, there is a wishy-washy sense in which organisms in a rainforest perform a valuable service for other species, and even for the maintenance of the whole forest community. Certainly, if you removed all the soil bacteria, the consequences for the trees and ultimately for most of the life of the forest, would be dire. But that is not why the soil bacteria are there. Yes, of course they do break down the dead leaves, dead animals and manure into compost which is useful for the continued prosperity of the whole forest. But they aren't doing it for the sake of making compost. They are using the dead leaves and dead animals as food for themselves, for the good of the genes that programme their compost-making activities. It is an incidental consequence of this self-interested activity that the soil is improved from the point of view of the plants, and therefore the herbivores that eat them, and therefore the carnivores that eat the herbivores. Species in a rainforest community flourish in the presence of the other species in that community because the community is the environment in which their ancestors survived. Maybe there are plants that flourish in the absence of a rich culture of soil bacteria, but those are not the plants we find in a rainforest. We are more likely to find them in a desert.

This is the right way to handle the temptation of 'Gaia': the overrated romantic fancy of the whole world as an organism; of each species doing its bit for the welfare of the whole; of bacteria, for example, working to improve the gas content of the earth's atmosphere for the good of all life. The most extreme example I know of this kind of bad poetic science comes from a famous and senior 'ecologist' (the quotation marks denote an activist for green politics, rather than a genuine scholar of the academic subject of ecology). It was reported to me by Professor John Maynard Smith, who was attending a conference sponsored by the Open University in Britain. The conversation turned to the mass extinction of the dinosaurs and whether this catastrophe was caused by a cometary collision. The bearded ecologist was in no doubt. 'Of course not,' he said decisively, '*Gaia would not have permitted it!*'

Gaia was the Greek earth goddess whose name has been adopted by James Lovelock, an English atmospheric chemist and inventor, to personify his poetic notion that the whole planet should be regarded as a single living thing. All living creatures are Gaia's body parts and they work together as a well-adjusted thermostat, reacting to perturbations so as to preserve all life. Lovelock is avowedly embarrassed by those, like the ecologist I have just quoted, who take his idea right over the top. Gaia has become a cult, almost a religion, and Lovelock now understandably wants to distance himself from this. But some of his own early suggestions, when you think them through, are only slightly more realistic. He proposed, for instance, that bacteria produce methane gas because of the valuable role it plays in regulating the chemistry of the earth's atmosphere.

The problem with this is that individual bacteria are asked to be nicer than natural selection can explain. The bacteria are supposed to produce methane beyond their own needs. They are expected to produce enough methane to benefit the planet in general. It is no good pleading that this is in their own long-term interests because if the planet goes extinct so will they. Natural selection is never aware of the long-term future. It is not aware of anything. Improvements come about not through foresight but by genes coming to outnumber their rivals in gene pools. Unfortunately, genes that make rebel bacteria sit back and enjoy the benefits of their rivals' altruistic production of methane are bound to prosper at the expense of the altruists. So the world will become relatively more full of selfish bacteria. This will continue even if, because of their selfishness, the total number of bacteria (and of everything else) is going down. It will continue even to the point of extinction. How should it not? There is no foresight.

If Lovelock were to retort that the bacteria produce methane as a by-product of something else that they do for their own good, and it is only incidentally useful for the world, I should agree wholeheartedly. But in that case the whole rhetoric of Gaia is superfluous and misleading. You don't need to talk about bacteria working for the good of anything other than their own short-term genetic good. We are left with the conclusion that individuals work for Gaia only when it suits them to do so – so why bother to bring Gaia into the discussion? We are better off thinking about genes, which are the real, self-replicating units of natural selection, flourishing in an environment which includes the genetic climate furnished by the other genes. I am quite happy to generalize the notion of the genetic climate to include all the genes in the whole world. But that is not Gaia. Gaia falsely focuses attention on planetary life as a single unit. Planetary life is a shifting pattern of genetic weather.

Lovelock's main comrade-in-arms as a champion of Gaia is the American bacteriologist Lynn Margulis. Despite her pugnacious disposition, she places herself firmly on the gentle side of the continuum which I am attacking as bad poetic science. Here she is, writing with her son Dorion Sagan:

> *Next, the view of evolution as chronic bloody competition among individuals and species, a popular distortion of Darwin's notion of 'survival of the fittest,' dissolves before a new view of continual cooperation, strong interaction, and mutual dependence among life forms. Life did not take over the globe by combat, but by networking. Life forms multiplied and complexified by co-opting others, not just by killing them.*
>
> *Microcosmos: Four Billion Years of Microbial Evolution* (1987)

Margulis and Sagan are in a superficial sense not too far from being right here. But they are misled by bad poetic science into expressing it wrongly. As I emphasized at the beginning of this chapter, the opposition 'combat versus cooperation' is the wrong dichotomy to stress. There is fundamental conflict at the level of the genes. But since the environments of genes are dominated by each other, cooperation and 'networking' arise automatically as a favoured manifestation of that conflict.

Where Lovelock is a student of the world's atmosphere, Margulis approaches from the other direction, as a specialist in bacteria. She rightly grants bacteria centre stage among life forms on our planet. At the level of biochemistry, there is a range of fundamental ways of making a living. These are practised by one or another kind of bacteria. One of these basic life recipes has been adopted by eukaryotes (that's everybody except bacteria), and we get it from bacteria. Margulis has successfully argued over many years that most of our biochemistry is carried out for us by what were once free bacteria now living within our cells. Here's another quotation from the same book by Margulis and Sagan.

Bacteria, by contrast, exhibit a far wider range of metabolic variations than eukaryotes. They indulge in bizarre fermentations, produce methane gas, 'eat' nitrogen gas right out of the air, derive energy from globules of sulfur, precipitate iron and manganese while breathing, combust hydrogen using oxygen to make water grow in boiling water and in salt brine, store energy by use of the purple pigment rhodopsin, and so forth . . . We, however, use just one of their many metabolic designs for energy production, namely that of aerobic respiration, the specialty of mitochondria.

Aerobic respiration, an elaborate set of biochemical cycles and chains whereby energy trapped from the sun is released from organic molecules, goes on in the mitochondria, the minute organelles that swarm inside our cells. Margulis has convinced the scientific world, rightly I think, that mitochondria are descended from bacteria. The ancestors of mitochondria, when they lived on their own, evolved the biochemical tricks that we call aerobic respiration. We eucaryotes now benefit from this advanced chemical wizardry because our cells contain the descendants of the bacteria that discovered them. On this view there is an unbroken line of descent from modern mitochondria back to ancestral bacteria living free in the sea. When I say 'line of descent', I literally mean that a free-living bacterial cell divided into two, and at least one of those divided into two, and at least one of those divided into two and so on until we reach every one of your mitochondria, continuing to divide in your cells.

Margulis believes that mitochondria were originally parasites (or predators – the distinction is not important at this level) which attacked the larger bacteria that were destined to provide the shell of the eucaryotic cell. There are still some bacterial parasites that do a similar trick, burrowing through the prey's cell wall, then, when safely inside, sealing up the wall and eating the cell from within. The mitochondrial ancestors, according to the theory, evolved from parasites that kill to less virulent parasites that keep their host alive to exploit it longer. Later still, the host cells began to benefit from the metabolic activities of the proto-mitochondria. The relationship shifted from predatory or parasitic (good for one side, bad for the other) to mutualistic (good for both). As the mutualism deepened, each came to depend more thoroughly on the other, and each came to lose those bits of itself whose purpose was best served by the other.

In a Darwinian world, such dedicatedly intimate cooperation evolves only when the DNA of the parasite passes 'longitudinally' down host generations in the same vehicles as the DNA of the host. To this day, our mitochondria still have their own DNA, which is only distantly related to our 'own' DNA and more closely related to that of certain bacteria. But it passes down human generations in human eggs. Parasites whose DNA passes longitudinally like this (that is, from host parent to host child) become less virulent and more cooperative, because anything that is good for the host DNA's survival automatically tends to be good for their own DNA's survival. Parasites whose DNA passes 'horizontally' (from a host to some other host which is not specifically its own child), for example rabies or flu viruses, may become even more virulent. If DNA is to be transmitted horizontally, the death of the host may be no bad thing. An extreme might be a parasite that feeds inside an individual host, turning its flesh into spores until it finally bursts, scattering the parasite DNA to the winds where it is blown far and wide to find new hosts.

Mitochondria are extreme longitudinal specialists. So intimate with the host cells did they become that it is hard for us to recognize that they were ever apart. My Oxford colleague Sir David Smith has found a neat simile:

In the cell habitat, an invading organism can progressively lose pieces of itself, slowly blending into the general background, its former existence betrayed only by some relic. Indeed, one is reminded of Alice in Wonderland's encounter with the Cheshire Cat. As she watched it, 'it vanished quite slowly, beginning with the tail, and ending with the grin, which remained some time after the rest of it had gone'. There are a number of objects in a cell like the grin of the Cheshire Cat. For those who try to trace their origin, the grin is challenging and truly enigmatic. *The Cell as a Habitat* (1979)

I don't find any strong distinction between the relationship of mitochondrial DNA to host DNA and that between one gene and another within the normal orthodox gene pool of a species' 'own' genes. I have argued that all our 'own' genes should be seen as mutually parasitic upon each other.

The other grinning relic which is now pretty uncontroversial is the chloroplast. Chloroplasts are small bodies in plant cells that do the business of photosynthesis – storing solar energy by using it to synthesize organic molecules. These organic molecules can then be broken down later and the energy released in a controlled way when required. Chloroplasts are responsible for the green colour of plants. It is now widely agreed that they are descended from photosynthetic bacteria, cousins of the 'blue-green' bacteria that still float free today and are responsible for 'blooms' in polluted water. The process of photosynthesis is the same in these bacteria and in (the chloroplasts of) eucaryotes. Chloroplasts, according to Margulis, were captured in a different way from mitochondria. Where the mitochondrial ancestors aggressively invaded larger hosts, the ancestors of chloroplasts were prey, originally engulfed for food, only later evolving a mutualistic rapport with their captors, doubtless again because their DNA became transmitted longitudinally down host generations.

More controversially, Margulis believes that yet another kind of bacteria, the spirally moving spirochaetes, invaded the early eucaryotic cell and contributed such moving structures as cilia, flagella and the 'spindles' which drag the chromosomes apart in cell division. Cilia and flagella (singular, cilium and flagellum) are just different-sized versions of each other, and Margulis prefers to call both 'undulipodia'. She reserves the name flagellum for the superficially similar but actually very different whip-like structure that some bacteria use to paddle ('screw' is a more appropriate verb) themselves along. The bacterial flagellum, incidentally, is remarkable in having the only true rotating bearing in the living kingdoms. It is nature's only important example of 'the wheel', or at least the axle, before humans reinvented it. Cilia and other eucaryotic undulipodia are more complicated. Margulis identifies each individual undulipodium with an entire spirochaete bacterium, in the same way as she identifies each mitochondrion and each chloroplast with an entire bacterium.

The idea of coopting bacteria to perform some difficult biochemical trick has frequently resurfaced in more recent evolution. Deep-sea fish have luminous organs to signal to each other and even to find their way about. Rather than undertake the difficult chemical task of making light, they have coopted bacteria that specialize in the skill. The luminous organ of a fish is a bag of carefully cultured bacteria, which give off light as a spin-off from their own biochemical purposes.

So we have a whole new way of looking at an individual organism.

Not only do animals and plants participate in complicated webs of interaction with each other, and with individuals of other species, in populations and communities like a tropical rainforest or a coral reef. Each individual animal or plant *is* a community. It is a community of billions of cells, and each one of those billions of cells is a community of thousands of bacteria. I'd go further and say that even a species' 'own' genes are a community of selfish cooperators. Now we are tempted by yet another piece of poetic science, the poetry of hierarchy. There are units within larger units, not just up to the level of the individual organism but even higher, for organisms live in communities. Is there not, at every level in the hierarchy, symbiotic cooperation between units at the next level down, units that used to be independent?

Perhaps there is some mileage in this. Termites make a very successful living out of eating wood and wood products such as books. But, yet again, the necessary chemical tricks do not come naturally to the termite's own cells. Just as the unaided eucaryote cell has to borrow the biochemical talents of mitochondria, so termite guts, on their own, cannot digest wood. They rely on symbiotic micro-organisms to do the business of digesting the wood. The termite itself subsists on the micro-organisms and their excreta. These micro-organisms are strange and specialized creatures, mostly found nowhere else in the world but in the guts of their own species of termite. They depend on the termites (to find the wood and chew it physically into small pieces), just as the termites depend on them (to break it down to even smaller molecular pieces, using enzymes which the termites themselves can't make). Some of the micro-organisms are bacteria, some of them are protozoa (single-celled eucaryotes), and some of them are a fascinating mixture of the two. Fascinating because of a kind of evolutionary *déjà vu* which powerfully adds plausibility to Margulis's speculation.

Mixotricha paradoxa is a flagellate protozoan which lives in the guts of the Australian termite *Mastotermes darwiniensis*. It has four large cilia at the front end. Margulis, of course, believes that these are themselves originally derived from symbiotic spirochaetes. But, though that may be controversial, there is a second kind of small, waving, hair-like projection about which there is no doubt. Covering the rest of the body, these look like cilia: cilia such as those that beat rhythmically to waft our eggs down our oviducts. But they are not cilia. Every one of them – and there are about half a million – is a tiny spirochaete bacterium. Indeed, there are two quite different kinds of spirochaete involved. It is these waving bacteria that propel *Mixotricha* around in the termite gut, and it is reported that they wave in unison. This seems hard to believe until you realize that each one could simply be provoked by its immediate neighbours.

The four large cilia at the front seem to serve only as rudders. These might be described as *Mixotricha*'s 'own', to distinguish them from the spirochaetes that carpet the rest of the body. But of course, if Margulis is right, they are really no more *Mixotricha*'s own than the spirochaetes: they just represent an earlier invasion. The *déjà vu* lies in the re-enactment by new spirochaetes, of a drama that was first staged a billion years ago. As it happens, *Mixotricha* cannot use oxygen because there isn't enough of it in the termite gut. Otherwise, we may be sure, they'd have mitochondria inside them – relic of yet another ancient wave of bacterial invasion. But in any case they *do* have other symbiotic bacteria inside them which are probably performing a biochemical role rather like mitochondria, maybe assisting with the difficult task of digesting wood.

A single *Mixotricha* individual, therefore, is a colony containing at least half a million symbiotic bacteria of various kinds. From a functional point of view as a wood-digesting machine, a single termite is a colony of perhaps as many symbiotic micro-organisms in its gut. Don't forget that, quite apart from the 'recent' invaders of its gut flora, a termite's 'own' cells, like the cells of any other eucaryote, are themselves colonies of much older bacteria. Finally, termites are rather special in that they themselves live in massive colonies of mostly sterile worker insects which plunder the country-side more effectively than almost any other kind of animal except ants – and they are successful for the same kind of reason. *Mastotermes* colonies can contain up to a million individual worker termites. The species is a voracious pest in Australia, devouring telegraph poles, the plastic lining of electric cables, wooden buildings and bridges, even billiard balls. Being a colony of colonies of colonies seems to be a successful recipe for life.

I want to return to the genes' eye view and push the idea of universal symbiosis – 'living together' – to its ultimate conclusion. Margulis is rightly seen as a high priestess of symbiosis. As I said earlier, I would go even further, and regard all 'normal' nuclear genes as symbiotic in the same kind of way as mitochondrial genes. But where Margulis and Lovelock invoke the poetry of cooperation and amity as primary in the union, I want to do the opposite and regard it as a secondary consequence. At the genetic level all is selfish, but the selfish ends of genes are served by cooperation at many levels. As far as the genes themselves are concerned, the relationships among our 'own' genes are not, in principle, different from the relationship between our genes and mitochondrial genes, or our genes and those of other species. All genes are being selected for their capacity to flourish in the presence of the other genes – of whatever species – whose consequences surround them.

Collaboration within gene pools to make complex bodies is often called co-adaptation, as distinct from co-evolution. Co-adaptation usually refers to the mutual tailoring of different bits of the same kind of organism to other bits. For example, many flowers have both a bright colour to attract insects and dark lines that act as runway guides to lead insects towards the nectar. Colour, lines and nectaries assist each other. They are co-adapted to each other, the genes that make them being selected in each other's presence. Co-evolution is normally used to mean mutual evolution in different species. The flowers and the insects that pollinate them evolve together – co-evolve. In this case the co-evolutionary relationship is mutually beneficial. The word co-evolution is also used for the hostile kind of evolving together – co-evolutionary 'arms races'. High-speed running in predators co-evolves with high-speed running in their prey. Thick armour co-evolves with weapons and techniques for penetrating it.

Although I have just made a clear distinction between 'within species' co-adaptation and 'between species' co-evolution, we can now see that a certain amount of confusion is pardonable. If we take the view, as I have in this chapter, that gene interactions are just gene interactions, at any level, co-adaptation turns out to be just a special case of co-evolution. As far as the genes themselves are concerned, 'within species' is not fundamentally different from 'between species'. The differences are practical. Within a species, genes meet their companions inside cells. Between species, their consequences in the outside world may meet the consequences of the other genes, out in the external world. Intermediate cases, such as intimate parasites and mitochondria, are revealing because they blur the distinction.

Sceptics of natural selection often worry along the following lines. Natural selection, they say, is a purely negative process. It weeds out the unfit. How can such a negative weeding-out play the *positive* role of building up complex adaptation? A large part of the answer lies in a combination of co-evolution and co-adaptation, two processes which, as we have just seen, are not so very far apart.

Co-evolution, like a human arms race, is a recipe for progressive build-up of improvements (I mean improvements in efficiency at what they do, of course; obviously, from a humane point of view, 'improvements' in armaments are just the reverse). If predators get better at their job, prey have to follow suit just to stay in the same place. And vice versa. The same goes for parasites and hosts. Escalation begets further escalation. This leads to real progressive improvement in equipment for survival, even if it does not lead to improvement in survival itself (because, after all, the other side in the arms race is improving too). So, co-evolution – arms races, the mutual evolution of genes in different gene pools – is one answer to the sceptic who thinks natural selection is a purely negative process.

The other answer is co-adaptation, the mutual evolution of genes in the *same* gene pool. In the cheetah gene pool, carnivorous teeth work best with carnivorous guts and carnivorous habits. Herbivorous teeth, guts and habits form an alternative complex in an antelope gene pool. At the gene level, as we have seen, selection puts together harmonious complexes, *not* by choosing whole complexes but by favouring each part of the complex within gene pools that are dominated by the other parts of the complex. In the shifting balance of gene pools, more than one stable solution to the same problem may exist. Once a gene pool starts to become dominated by one stable solution, further selection of selfish genes favours the ingredients of the same solution. The other solution could equally well have been favoured if the starting conditions had been different. In any case, the sceptic's worry about whether natural selection is purely a negative, subtracting process is disarmed. Natural selection is positive and constructive. It is no more negative than a sculptor subtracting marble from a block. It carves out of gene pools complexes of mutually interacting, co-adapted genes: fundamentally selfish but pragmatically cooperating. The unit that the Darwinian sculptor carves is the gene pool of a species.

I have devoted space in the last couple of chapters to warning of bad poetry in science. But the balance of my book is the opposite. Science is poetic, ought to be poetic, has much to learn from poets and should press good poetic imagery and metaphor into its inspirational service. 'The selfish gene' is a metaphoric image, potentially a good one but capable of sadly misleading if the metaphor of personification is improperly grasped. If interpreted aright it may lead us into paths of deep understanding and fertile research. This chapter has used the metaphor of the personified gene to explain a sense in which 'selfish' genes are also 'cooperative'.

The key image to be floated in the next chapter is that of a species' genes as a detailed *description* of the collection of environments in which its ancestors lived – a genetic book of the dead.

10

THE GENETIC BOOK OF THE DEAD

Remember the wisdom out of the old days . . .

W. B. YEATS,
The Wind Among the Reeds (1899)

The first essay I can remember writing at school was 'The Diary of a Penny'. You had to imagine yourself a coin and tell your story, of how you sat in a bank for a while until you were given out to a customer, how it felt to jangle around in his pocket with the other coins, how you were handed over to buy something, then how you were passed out as change to another customer and then . . . well, you probably wrote a similar essay yourself. It is helpful to think the same way for a gene travelling not from pocket to pocket but from body to body down the generations. And the first point that the analogy of the coin makes is that *of course* the personification of the gene is not to be taken literally, any more than we seven-year-olds really thought our coins could talk. Personification is sometimes a useful device, and for critics to accuse us of taking it literally is almost as stupid as taking it literally in the first place. Physicists are not literally charmed by their particles, and the critic who would so accuse them is a tiresome pedant.

The 'minting' event for a gene is the mutation that brought it into existence by altering a previous gene. Only one of many copies of the gene in the population is changed (by one mutation event, but an identical mutation may change another copy of the gene in the gene pool at another time). The others continue to make copies of the original gene, which may now be said to be in competition with the mutant form. Making copies is of course what genes, unlike coins, are supremely good at, and our diary of a gene has to include the experiences not of the particular atoms that go to make up the DNA, but the DNA's experiences in the form of multiple copies in successive generations. As the last chapter showed, much of a gene's 'experience' of past generations consists of rubbing up against other genes of the species, and this is why they cooperate so amicably in the collective enterprise of building bodies.

Now let's ask the question whether all the genes of a species have the same ancestral 'experiences'. Mostly they do. Most buffalo genes can look back to a long line of buffalo bodies which enjoyed, or suffered, common buffalo experiences. The bodies in which these genes survived included male and female buffaloes, large and small, and so on. But there are subsets of genes with different experiences, for example the genes that determine sex. In mammals, Y chromosomes are found only in males and do not exchange genes with other chromosomes. So a gene sitting on a Y chromosome has had a limited experience of buffalo bodies: male ones only. Its experiences are *largely* typical of buffalo genes in general, but not entirely so. Unlike most buffalo genes it doesn't know what it is like to sit in a female buffalo. A gene that has always been on a Y chromosome since the origin of the mammals during the age of dinosaurs will have experienced male bodies of many different species, but never a female body of any kind. The case of X chromosomes is more complicated to work out. Male mammals have one X chromosome (inherited from the mother, plus one Y chromosome inherited from the father), while females have two X chromosomes (one from each parent). So each X chromosome gene has experienced both female and male bodies, but two-thirds of its experience has been in female bodies. In birds the situation is reversed. The female bird has uneven sex chromosomes (which we may as well call X and Y by analogy with mammals, although the official bird terminology is different), the male two of the same (XX).

The genes on the other chromosomes have all had an equal experience of male and female bodies, but their experiences may be unequal in other respects. A gene will have spent more than its fair share of time in ancestral bodies that possess whatever qualities the gene encodes – long legs, thick horns, or whatever it may be, especially if it is a dominant gene. Almost as obviously, all genes are likely to have spent more of their ancestral time in successful than in unsuccessful bodies. There are plenty of unsuccessful bodies about and they contain their full complement of genes. But they tend not to have descendants (that is what being unsuccessful means) so, as a gene looks back through its biography of past bodies, it will observe that all of them were as a matter of fact successful (by definition), and perhaps most (but not all) of them were equipped with what it normally takes to be successful. The difference here is that individuals who are not equipped to be successful sometimes reproduce despite their lack. And individuals who are superbly equipped to survive and reproduce under average conditions are sometimes struck by lightning.

If, like some deer, seals and monkeys, the species is one in which the males form dominance hierarchies and dominant males do most of the reproducing, it will follow that the genes of the species will have more experience of dominant male bodies than of subordinate ones. (Note that we are no longer using dominant in its technical genetic sense, whose opposite is recessive, but in its ordinary language sense, where the opposite is subordinate.) In every generation, most of the males are subordinate, but their genes still look back on a strong line of dominant male ancestors. In every generation, the majority are fathered by a dominant minority from the previous generation. In the same way, if, like pheasants, the species is one in which, we suppose, most of the insemination is done by beautiful (to the females) males, most genes, whether they are now in females, in ugly males or in beautiful males, can look back on a long line of beautiful male ancestors. Genes have more experience of successful bodies than of unsuccessful ones.

To the extent that the genes of a species have regular and recurrent experience of subordinate bodies, we can expect to witness conditional strategies for 'making the best of a bad job'. In those species where successful males pugnaciously defend large harems, we sometimes notice subordinate males employing alternative, 'sneaky' strategies for gaining fleeting access to females. Seals have some of the most harem-dominated societies in the animal kingdom. In some populations, more than 90 per cent of the copulations are achieved by fewer than 10 per cent of the males. The bachelor majority of males, while biding their time awaiting their moment to depose one of the harem-bossing bulls, are alert for opportunities to sneak copulations with temporarily unguarded females. But, for such an alternative male strategy to have been favoured by natural selection, there must be at least a significant trickle of genes that have sneaked down the generations via stolen copulations. In our 'diary of a gene' language, then, at least some genes record subordinate males in their ancestral experience.

Do not be misled by the word 'experience'. It is not just that the word must be taken metaphorically rather than literally. That, I hope, is obvious. Less obvious is that we get a much more telling metaphor if we think of the whole species' gene pool, rather than a single gene, as the entity that gains experience from its ancestral past. This is another aspect of our doctrine of 'the selfish cooperator'. Let me try and spell out what it means to say of a species, or its gene pool, that it learns from its experiences. The species changes over evolutionary time. In any one generation, of course, the species consists of the set of its individual members alive at that time. Obviously this set changes as new members are born and old members die. This change in itself does not deserve to be called benefiting from experience, but the statistical distribution of genes in the population may systematically move in some specified direction, and that is 'species experience'. If an ice age is creeping up, more and more individuals will be seen to have thick hairy coats. Those individuals that happen to be the hairiest in any one generation tend to contribute more than their fair share of offspring, and hence genes for hairiness, to the next generation. The set of genes in the whole population – and therefore the genes likely to be contained in a typical average individual – becomes shifted in the direction of more and more genes for hairiness. The same thing is going on for other kinds of genes. As the generations go by, the whole set of genes of a species – the gene pool – is carved and whittled, kneaded and shaped, so that it becomes good at making successful individuals. It is in this sense that I say that the species is learning from its experience in the art of building good individual bodies, and it stores its experiences in coded form in the set of genes in the gene pool. Geological time is the timescale over which species become experienced. The information that the experience packs away is information about ancestral environments and how to survive them.

A species is an averaging computer. It builds up, over the generations, a statistical description of the worlds in which the ancestors of today's species members lived and reproduced. That description is written in the language of DNA. It lies not in the DNA of any one individual but collectively in the DNA – the selfish cooperators – of the whole breeding population. Perhaps 'read-out' captures it better than 'description'. If you find an animal's body, a new species previously unknown to science, a knowledgeable zoologist allowed to examine and dissect its every detail should be able to 'read' its body and tell you what kind of environment its ancestors inhabited: desert, rainforest, arctic tundra, temperate woodland or coral reef. The zoologist should also be able to tell you, by reading its teeth and its guts, what it fed on. Flat, millstone teeth and long intestines with complicated blind alleys indicate that it was a herbivore; sharp, shearing teeth and short, uncomplicated guts indicate a carnivore. The animal's feet, its eyes and other sense organs spell out the way it moved and how it found its food. Its stripes or flashes, its horns, antlers or crests, provide a read-out, for the knowledgeable, of its social and sex life.

But zoological science has a long way to go. Present day zoology can 'read' the body of a newly discovered species only to the extent of a rough, qualitative verdict about its probable habitat and way of life. The zoology of the future will put into the computer many more measurements of the anatomy and chemistry of the animal being 'read'. More importantly, we shall not take the measurements separately. We shall perfect mathematical techniques of combining information from teeth, guts, stomach chemistry, social coloration and weapons, blood, bones, muscles and ligaments. We shall incorporate methods of analysing the interactions of these measurements with one another. The computer, combining everything that is known about the body of the strange animal, will construct a detailed, quantitative model of the world, or worlds, in which the animal's ancestors survived. This, it seems to me, is tantamount to saying that the animal, any animal, *is* a model or description of its own world, or more precisely the worlds in which its ancestors' genes were naturally selected.

In a few cases, an animal's body is a description of the world in the literal sense of a pictorial representation. A stick insect lives in a world of twigs, and its body is a representational sculpture of a twig, leaf scars, buds and all. A fawn's pelage is a painting of the dappled pattern of sunlight filtered through trees on to the woodland floor. A peppered moth is a model of lichen on the tree bark. But just as art doesn't have to be literalist and representational, animals can be said to render their world in other ways: impressionistic, say, or symbolic. An artist seeking a dramatic impression of air speed could hardly do better than the shape of a swift. Perhaps this is because we have an intuitive understanding of streamlining; perhaps it is because we have grown used to the swept-back beauty of modern jet planes; perhaps it is because we have picked up some knowledge of the physics of turbulence and Reynolds Numbers, in which case we could say that the shape of the swift embodies coded facts about the viscosity of the air in which its ancestors flew. Whichever is the case, we see a swift as fitting the world of high-speed airflow as a hand fits a glove, an impression enhanced when we contrast it with the floundering clumsiness of a swift stranded on the ground and unable to take off.

A mole is not literally the shape of an underground tunnel. Perhaps it is a kind of negative image of a tunnel, shaped to squeeze through it. Its hands are not literally like soil, but they resemble spades which, through experience or intuition, we can see as the functional complement of soil: spades powered by heavy muscles to work against soil. There are even more striking cases where an animal, or a part of an animal, does not literally resemble its world but fits some part of it, glove-fashion. The coiled abdomen of a hermit crab is a coded representation of the mollusc shells in which its ancestors' genes lived. Or, we could say that the hermit crab's genes contain a coded prediction about an aspect of the world in which the crab will find itself. Because modern snails and whelks are on average the same as ancient snails and whelks, hermit crabs still fit them and survive – the prediction is fulfilled.

Species of tiny mite are specialized to ride at a precise location on the inside of the pincer-like mandibles of a particular caste of army ant workers. Another species of mite is specialized to ride on the first joint of one antenna of an army ant. Each of these mites is shaped to fit its precise habitat, as a key fits a lock (Professor C. W. Rettenmeyer informs me – to my regret – that there are not mites shaped for left antennae and other mites shaped for right antennae). Just as a key embodies (complementary or negative) information about its own lock (information without which the door cannot be opened), so the mite embodies information about its world, in this case the shape of the insect joint where it lodges. (Parasites are often very specialized keys which fit their hosts' locks in much more detail than predators, presumably because it is unusual for a predator to attack only one species of host. The distinguished biologist Miriam Rothschild has a fund of delightful examples including a 'worm which lives exclusively under the eyelids of the hippopotamus and feeds upon its tears'.)

Sometimes the fit of animal to world is intuitively clear, either to common sense or to the trained eye of the engineer. Anybody can work out why webbed feet are so common among animals that frequently enter the water – ducks, platypuses, frogs, otters and others. If you are in any doubt, put on a pair of rubber frog feet and experience the sense of immediate release when you swim. You might even wish you had been born with frog feet, until you get out of the water and try to walk in your rubber ones. My friend Richard Leakey, paleoanthropologist, conservationist and African hero, lost both his legs in a light aircraft crash. Now he has two pairs of artificial legs: one pair with shoes, extra large for stability and permanently laced for walking, and another pair with flippers for swimming. Feet that are good for one way of life are bad for another. It is hard to design an animal that can do two such different things well.

Anybody can see why otters, seals and other air-breathing animals that live in water often have nostrils that can be closed at will. Again, human swimmers often resort to artifice, in this case a sprung nose-clip like a clothes peg. Anybody watching an anteater feeding through a hole in a nest of ants or termites can quickly see why they have been furnished with a long thin snout and sticky tongue. This is true not just of the specialized anteaters of South America, but of the unrelated pangolins and aardvark of Africa, and the even less closely related numbat and very distantly related spiny anteaters of Australasia. It is less obvious why all mammals that eat ants or termites have a low metabolic rate – a low body temperature compared with most mammals, and a correspondingly low rate of biochemical turnover.

Our zoologists of the future, in order to reconstruct ancestral worlds and genetic descriptions of them, will need to replace intuitive common sense with systematic research. Here's how they might proceed. Begin by listing a set of animals which are not particularly closely related to each other but which all share an important aspect of their life. Water-dwelling mammals would be a good test case. On more than a dozen separate occasions, land-dwelling mammals have returned to make their living, either wholly or partly, in water. We know they did so independently of each other because their closer cousins still live on land. The Pyrenean desman is a kind of aquatic mole, closely related to our familiar burrowing moles. Desmans and moles are members of the order *Insectivora*. Other members of the *Insectivora* who have independently evolved to live in fresh water include water shrews, one aquatic species of the exclusively Madagascan group of tenrecs, and three species of related otter shrews. That's four separate returnings to water among the *Insectivora* alone. All four are closer cousins to relatives living on dry land than they are to the other freshwater species in the list. We have to count the three otter shrews as only a single returning to water because they are related to each other and presumably are all descended from a recent aquatic ancestor.

The surviving whales probably represent at most two separate returnings to water: the toothed whales (including dolphins) and the baleen whales. The surviving dugongs and manatees are close cousins of each other, and certainly their common ancestor also lived in the sea: they too, then, represent only a single returning to the sea. Within the pig family, most live on land but hippopotamuses have partly returned to live in water. Beavers and true otters are other animals whose ancestors returned to the water. They can be directly compared with cousins who have stayed on land, say prairie dogs in the case of beavers, and badgers in the case of otters. Mink are members of the same genus as weasels and stoats (that puts them as close to each other as horses, zebras and donkeys are to each other) but they are semi-aquatic and have partially webbed feet. There is a South American water marsupial, the yapok, which can be directly compared with its land-dwelling cousins among the opossums. Among the egg-laying mammals of Australasia, the duck-billed platypuses live largely in water, the spiny anteaters on land. We can make a respectable list of matched pairs: each independently evolved aquatic group opposite the closest cousin we can find that has stayed on the land.

Given the list of matched pairs, we can immediately notice some obvious things. Most of the water-dwellers have at least partially webbed feet; some have a tail that is modified into the shape of a paddle. These are obvious in the same kind of way as the long sticky tongue shared by anteaters. But, like the low metabolic rate shared by anteaters, there are probably less obvious characteristics shared by aquatic mammals as distinct from their terrestrial cousins. How shall we discover them? By a systematic statistical analysis; perhaps something like this.

Looking down our table of matched pairs, we make a large set of measurements, the same measurements for all the animals. We measure anything we can think of, with no prior expectations: pelvis width, eye radius, gut length, dozens more, all perhaps scaled as a ratio of total body size. Now throw all the measurements into the computer and invite it to work out which measurements need to be given high *weighting* in order to discriminate aquatic animals from their terrestrial cousins. We could calculate a number, the 'discrimination number', by summing up contributions from all the measurements, each one having been multiplied by a weighting factor. The computer adjusts the weighting given to each measurement, in order to maximize the difference, in the final sum, between aquatic mammals and their terrestrial opposite numbers. The foot webbing index will presumably emerge from the analysis with a strong weighting. The computer will discover that it pays – if you are trying to maximize the difference between aquatic and terrestrial animals – to multiply the webbing index by a high number before adding it to the discrimination number. Other measurements – things that mammals share without regard to the wetness of their world – will need to be multiplied by zero in order to eliminate their irrelevant and confusing contribution to the weighted sum.

At the end of the analysis we look down the weightings of all our measurements. Those that emerge with high weightings, like the foot webbing index, are the ones that have something to do with the wateriness of the way of life. Webbiness is obvious. What we hope is that the analysis will uncover other important discriminators that are not so obvious. Biochemical measures, for instance. When we have found them, we can then scratch our heads and wonder what connection they have with living in water or on land. This may suggest hypotheses for further research. Even if it doesn't, any measurement that gives us a statistically significant difference between animals that have adopted some way of life and their cousins that have not is very likely to be telling us something important about that way of life.

We can do the same thing with genes. Without any prior hypotheses about what the genes are doing, we make a systematic search for genetic resemblances between unrelated aquatic animals, which are not shared by their terrestrial close cousins. If we find any strong and statistically significant effects, even if we don't understand what those genes are doing, I would say that what we are looking at might be regarded as a genetic description of watery worlds. Natural selection, to repeat, works as an averaging computer, doing the equivalent of a calculation that is not unlike the calculations we have just programmed our manmade computer to perform.

Often a species embraces various ways of life, which may be radically different from each other. A caterpillar, and the butterfly it becomes, are members of the same species, yet our zoologist's reconstruction of their two ways of life would be utterly different. Caterpillar and butterfly contain exactly the same set of genes, and the genes must describe both environments, but separately. Presumably many of them are turned on in the plant-chewing, growing, caterpillar phase, and a largely different set of genes are turned on in the adult, reproductive, nectar-eating phase.

Male and female of most species live in at least somewhat different ways. The differences are pushed to extremes in angler fish, where the male appends himself as a tiny parasitic protuberance on the female's massive body. In most species, including ourselves, both male and female contain most of the genes for being either male or female. The differences lie in which genes are turned on. We all have genes for making penises and genes for making uteruses, regardless of our sex. ('Sex' is correct, by the way, not 'gender'. Gender is a grammatical technical term, applied to words not creatures. In German, a girl's gender is neuter but her sex female. Amerindian languages typically have two genders, animate and inanimate. The association of gender with sex in some groups of languages is incidental. It is quite a good joke that the politically inspired euphemism – saying gender when you mean sex – is consequently a piece of Western imperialism.) Our future zoologist reading the body of either a male or a female would get an incomplete picture of the ancestral worlds of the species. On the other hand, the *genes* of any member of the species would more nearly suffice to reconstruct a complete picture of the range of ways of life that the species has experienced.

Parasitic cuckoos are an oddity, and a fascinating one from the point of view of the Genetic Book of the Dead. As is well known, they are reared by foster parents of a species not their own. They never rear their own young. Not all are reared by the same foster species. In Britain, some are reared by meadow pipits, some by reed warblers, fewer by robins, some by a variety of other species, but the largest number are reared by dunnocks (hedge sparrows). As it happens, our foremost specialist on dunnocks and the author of *Dunnock Behaviour and Social Evolution* (1992) is also today's leading investigator of cuckoo biology, Nicholas Davies of Cambridge University. I shall base my account on the work of Davies and his colleague Michael Brooke because it lends itself especially well to being cast in the language of species 'experience' of ancestral worlds. Unless otherwise stated I shall refer to the common cuckoo, *Cuculus canorus*, in Britain.

Although they make mistakes on 10 per cent of occasions, a female cuckoo normally lays in the same kind of nest as her mother, her maternal grandmother, her maternal maternal great grandmother, and so on. Presumably young females learn the characteristics of their foster nest and seek it out when their own time comes to lay. So, as far as females are concerned, there are dunnock cuckoos, reed warbler cuckoos, meadow pipit cuckoos, and so on, and they share this attribute with their female-line relatives. But these are not separate species, not even separate races in the normal sense of the word. They are called 'gentes' (singular 'gens'). The reason a gens is not a true race or species is that male cuckoos don't belong to a gens. Since males don't lay eggs, they never have to choose a foster nest. And when a male cuckoo comes to mate, he just mates with a female cuckoo irrespective of her gens, and regardless of the foster species that reared either of them. It would follow from this that there is gene flow between the gentes. Males carry genes from one female gens to another. A female's mother, maternal grandmother and maternal maternal great grandmother will all belong to the same gens. But her paternal grandmother, both her paternal great grandmothers, and all her female ancestors to whom she is linked via any male ancestor, could belong to any gens. From the point of view of gene 'experience', the consequence is very interesting. Recall that, in birds, it is the female sex that has the unequal sex chromosomes, X and Y, while male birds have two X chromosomes. Think what this means for the ancestral experience of genes on a Y chromosome. Since it passes unswervingly down the female line, never straying into the paths of male experience, a Y chromosome stays strictly within one gens. It is a dunnock cuckoo Y chromosome or a meadow pipit cuckoo Y chromosome. Its foster parent 'experience' is the same from generation to generation. In this respect it differs from all other genes in the cuckoo, for they have all done time in male bodies and have hence shuffled freely around the female gentes, experiencing them all in proportion to their frequency.

In our language of genes as 'descriptions' of ancestral environments, most cuckoo genes will be in a position to describe those features that are shared by the complete range of foster nests that the species has parasitized. Y chromosome genes, uniquely, will describe only one type of foster nest, one species of foster parent. This means that Y chromosome genes, in a way that is not possible for other cuckoo genes, will be in a position to evolve specialized tricks for surviving in their own particular foster species' nest. What sort of tricks? Well, cuckoo eggs show at least some tendency to mimic the eggs of their foster species. Cuckoo eggs laid in meadow pipit nests are like large meadow pipit eggs. Cuckoo eggs laid in reed warbler nests are like large reed warbler eggs. Cuckoo eggs laid in pied wagtail nests resemble pied wagtail eggs. Presumably this benefits the cuckoo eggs, which might otherwise be rejected by the foster parents. But think what it must mean from the genes' point of view.

If the genes for egg colour were on any chromosomes but the Y chromosome, they would be carried via males into the bodies of females belonging to the full assortment of gentes. This means they would be carried into the full range of host nests and there would be no consistent natural selection pressure to mimic one egg type more than another. It would be difficult for their eggs, in these circumstances, to mimic any but the most generalized features of all host eggs. Although there is no direct evidence, it is therefore reasonable to guess that the specific egg mimicry genes sit on the cuckoo Y chromosome. Females will then carry them, generation after generation, into nests of the same host. Their ancestral 'experience' will all be with the discriminating eyes of the same host, and those eyes will exert the selection pressure that steers their colour and spot pattern towards mimicry of host eggs.

There is a conspicuous exception. Cuckoo eggs laid in dunnock nests do not resemble dunnock eggs. They are no more variable among themselves than eggs laid in reed warbler or meadow pipit nests; their colour is distinctive of the dunnock gens of cuckoos, and they don't much resemble the eggs of any other gens, but they don't resemble dunnock eggs either. Why is this? It might be thought that dunnock eggs, being a uniform pale blue, are harder to mimic than meadow pipit or reed warbler eggs. Perhaps cuckoos just lack the physiological equipment to make plain blue eggs? I'm always suspicious of such last-resort theories, and in this case there is evidence against it. In Finland there is a gens of cuckoo that parasitizes redstarts, which also have plain blue eggs. These cuckoos, which belong to the same species as our British ones, succeed beautifully in matching redstart eggs. This surely shows that the failure of British cuckoos to mimic dunnock eggs cannot be put down to inherent incapacity to produce the unspeckled blue colour.

Davies and Brooke believe that the true explanation lies in the recency of the relationship between dunnocks and cuckoos. Cuckoos run arms races in evolutionary time with each host species and the gens we are looking at has only recently 'invaded' dunnocks. Consequently, dunnocks haven't yet had time to evolve counter-weapons. And dunnock cuckoos either haven't had time to evolve eggs that mimic dunnocks, or they don't yet need to because dunnocks haven't evolved the habit of discriminating foreign eggs from their own. In the language of this chapter, neither the dunnock gene pool nor the cuckoo gene pool (or rather, the Y chromosome of the dunnock cuckoo gens) has had enough experience of the other to evolve counter-weapons. Perhaps dunnock cuckoos are still adapted to outwit a different foster species, the one that their female ancestor left when she laid the first egg in a dunnock nest.

Meadow pipits, reed warblers and pied wagtails, on this view, are old enemies of their respective gentes of cuckoos. There has been plenty of time for a build-up of weaponry on both sides. The hosts have built up keen eyes for an impostor egg and the cuckoos possess correspondingly cunning disguises for their eggs. Robins are an intermediate case. Their cuckoos lay eggs which are slightly robin-like, but not very. Perhaps the arms race between robins and the robin gens of cuckoos is of intermediate antiquity. On this view, the Y chromosomes of robin cuckoos are somewhat experienced, but their description of recent (robin) ancestral environments is still sketchy and contaminated by earlier descriptions of other species, previously 'experienced'.

Davies and Brooke did experiments deliberately putting extra eggs, of various kinds, in nests belonging to different species of birds. They wanted to see which species would accept, or reject, strange eggs. Their hypothesis was that species that have been through an arms race with cuckoos would, as a consequence of their genetic 'experience', be most likely to reject foreign eggs. One way to test this was to look at species which are not even suitable as cuckoo hosts. Baby cuckoos need to eat insects or worms. Species that feed their young on seeds, or species that nest in holes that female cuckoos can't reach, have never been at risk. Davies and Brooke predicted that such birds would not worry if they experimentally introduced strange eggs into their nests. And so it proved. Species that are suitable for cuckoos, however, like chaffinches, song thrushes and blackbirds, showed a stronger tendency to reject the experimental eggs that Davies and Brooke, playing cuckoo, placed in their nests. Flycatchers are potentially vulnerable in that they feed their young on a cuckoo-friendly diet. But whereas spotted flycatchers have open and accessible nests, pied flycatchers nest in holes which female cuckoos are too large to penetrate. Sure enough, when the experimenters dumped foreign eggs in their nests, pied flycatchers, with their 'inexperienced' gene pools, accepted foreign eggs without protest; spotted flycatchers, by contrast, rejected them, suggesting that their gene pools were wise to the cuckoo menace from long ago.

Davies and Brooke did similar experiments with species that cuckoos actually do parasitize. Meadow pipits, reed warblers and pied wagtails usually rejected artificially added eggs. As befits the 'lack of ancestral experience' hypothesis, dunnocks did not; nor did wrens. Robins and sedge warblers were intermediate. At the other extreme, reed buntings, which are suitable for cuckoos but not much parasitized by them, showed total rejection of foreign eggs. No wonder cuckoos don't parasitize them. Davies and Brooke's interpretation would presumably be that reed buntings have come out the other side of a long ancestral arms race with cuckoos, which they eventually won. Dunnocks are near the beginning of their arms race. Robins are slightly more advanced in theirs. Meadow pipits, reed warblers and pied wagtails are in the middle of theirs.

When we say dunnocks have only just begun their arms race with cuckoos, 'only just' has to be interpreted with evolutionary timescales in mind. By human standards the association could still be quite old. The *Oxford English Dictionary* quotes a 1616 reference to the *Heisugge* (archaic word for hedge sparrow or dunnock) as 'a bird which hatcheth the Cuckooes egges'. Davies notes the following lines in *King Lear* I, iv, written a decade earlier:

> *For, you trow, nuncle,*
> *The hedge-sparrow fed the cuckoo so long,*
> *That it's had it head bit off by it young.*

And in the fourteenth century Chaucer wrote of the cuckoo's treatment of the dunnock in *The Parliament of Fowls:*

> *'Thou mortherere of the heysoge on the braunche*
> *That broughte the forth, thow rewthelees glotoun!'*

Although dunnock, hedge sparrow and heysoge are all given as synonyms in the dictionary, I can't help wondering how far we should rely on medieval ornithology. Chaucer himself was usually a rather precise user of language, but nevertheless the name sparrow has at times been given to what today is technically called an LBB (little brown bird). This may have been Shakespeare's meaning in the following, from *Henry IV Part 1*, V, i:

> *And, being fed by us, you used us so*
> *As that ungentle gull, the cuckoo's bird,*
> *Useth the sparrow – did oppress our nest;*

Grew by our feeding to so great a bulk
That even our love durst not come near your sight
For fear of swallowing;

Sparrow, on its own, would nowadays mean the house sparrow, *Passer domesticus*, which is never parasitized by cuckoos. Despite its alternative name hedge sparrow, the dunnock, *Prunella modularis*, is unrelated; it is a 'sparrow' only in the loose sense of being a little brown bird. But anyway, even if we take Chaucer's evidence as showing that the arms race between cuckoos and dunnocks really does go back at least to the fourteenth century, Davies and Brooke cite theoretical calculations, taking into account the comparative rarity of cuckoos, suggesting that this is still sufficiently recent in evolutionary terms to account for the apparent naivety of dunnocks when faced with cuckoos.

Before we leave cuckoos, here's an interesting thought. There could be, simultaneously existing, more than one gens of, say, robin cuckoos, who have built up their egg mimicry independently. Since there is no gene flow between them as far as Y chromosomes are concerned, there could be accurate egg mimics coexisting with less accurate egg mimics. All are capable of mating with the same males but they don't share the same Y chromosomes. The accurate mimics would be descended from a female who moved into parasitizing robins a long time ago. The less accurate ones would be descended from a different female who moved into robins, possibly from a different predecessor host species, more recently.

Ants, termites and other social insect species are odd in a different way. They have sterile workers, often divided into several 'castes' – soldiers, media (middle-sized) workers, minor (small) workers, and so on. Every worker, whatever its caste, contains the genes that could have turned it into any other caste. Different sets of genes are switched on under different rearing conditions. It is by regulating these rearing conditions that the colony engineers a useful balance of different castes. Often the differences among castes are dramatic.

In the Asian ant species *Pheidologeton diversus*, the large worker caste (specialized for bulldozing smooth paths for other colony members) is 500 times heavier than the small caste, who do all the normal duties of a worker ant. The same set of genes equips a larva to grow up into either a Brobdingnagian or a Lilliputian, depending upon which ones are switched on. Honeypot ants are immobile storage vats, abdomens pumped up with nectar to transparent yellow spheres, hanging from the ceiling of the nest. The normal duties of an ants' nest, defence, foraging and, in this case, filling up the living vats, are done by normal workers whose abdomens are not swollen. The normal workers have genes that equip them to be honeypots, and honeypots, as far as their genes are concerned, could equally well be normal workers. As in the case of male and female, the visible differences in bodily form depend upon which genes are switched on. In this case it is determined by environmental factors, perhaps diet. Once again, the zoologist of the future could read out from the genes, but not the body, of any one member of the species a complete picture of the disparate lives of the different castes.

The European snail *Cepaea nemoralis* comes in a number of colours and patterns. The background shell colour can be any of six distinct shades (in order of dominance, in the technical genetic sense): brown, dark pink, light pink, very pale pink, dark yellow, light yellow. Overlaying this, there may be any number of stripes from zero to five. Unlike the case of the social insects, it is not true that every individual snail is genetically equipped to assume any of the different forms. Nor are these differences among snails determined by different environments of upbringing. Striped snails have genes that determine their number of stripes, dark pink individuals have genes that make them dark pink. But all the kinds can mate with each other.

The reasons for the persistence of many different types of snail (polymorphism), as well as the detailed genetics of the polymorphism itself, have been exhaustively studied by the English zoologists A. J. Cain and the late P. M. Sheppard with their school. A major part of the evolutionary explanation is that the species ranges over different habitats – woodland, grassland, bare soil – and you need a different colour pattern to be camouflaged against birds in each place. Beechwood snails contain an admixture of genes from grassland because they interbreed at the margins. A chalk downland snail has some genes that previously survived in the bodies of woodland ancestors; and their legacy, depending on the other genes in the snail, may be stripes. Our zoologist of the future would need to look at the gene pool of the species as a whole to reconstruct the full range of its ancestral worlds.

Just as *Cepaea* snails range over different habitats in space, so the ancestors of any species have changed their way of life from time to time. House mice, *Mus musculus*, today live almost exclusively in or around human habitations, as unwanted beneficiaries of human agriculture. But by evolutionary standards their way of life is recent. They must have fed on something else before there was human agriculture. Doubtless that something was sufficiently similar for their genetic skills to be pressed into service when the agricultural bonanza came along. Mice and rats have been described as animal weeds (incidentally, a good piece of poetic imagery, genuinely illuminating). They are generalists, opportunists, carrying genes that helped their ancestors to survive through probably a considerable range of ways of life; and pre-agricultural genes are in them yet. Anybody attempting to 'read' their genes may find a confusing palimpsest of ancestral world descriptions.

From earlier still, the DNA of all mammals must describe aspects of very ancient environments as well as more recent ones. The DNA of a camel was once in the sea, but it hasn't been there for a good 300 million years. It has spent most of recent geological history in deserts, programming bodies to withstand dust and conserve water. Like sandbluffs carved into fantastic shapes by the desert winds, like rocks shaped by ocean waves, camel DNA has been sculpted by survival in ancient deserts, and even more ancient seas, to yield modern camels. Camel DNA speaks – if only we could understand the language – of the changing worlds of camel ancestors. If only we could read the language, the DNA of tuna and starfish would have 'sea' written into the text. The DNA of moles and earthworms would spell 'underground'. Of course all the DNA would spell many other things as well. Shark and cheetah DNA would spell 'hunt', as well as separate messages about sea and land. Monkey and cheetah DNA would spell 'milk'. Monkey and sloth DNA would spell 'trees'. Whale and dugong DNA presumably describes very ancient seas, fairly ancient lands and more recent seas: complicated palimpsests again.

Features of the environment that occur frequently or importantly are heavily emphasized or 'weighted' in the genetic description, compared with rare or trivial features. Environments that lie in the remote past have a different weighting from recent ones, presumably lower, though not in any obvious way. Environments that lasted a long time in the species' history will have a more prominent weighting in the genetic description than environmental events that, however drastic they may have seemed at the time, were geological flashes in the pan.

It has been poetically suggested that the remote marine apprenticeship of all land life is reflected in the biochemistry of the blood, which is said to resemble a primeval salt sea. Or the liquid in a reptile's egg has been described as a private pond, relic of the actual ponds in which the larvae of distant, amphibious ancestors would have grown. To the extent that animals and their genes bear such a stamp of ancient history it will be for good functional reasons. It won't be history for history's sake. Here is the kind of thing I mean by this. When our remote ancestors lived in the sea, many of our biochemical and metabolic processes became geared to the chemistry of the sea – and our genes became a description of marine chemistry – for functional reasons. But (this is an aspect of our 'selfish cooperator' argument) biochemical processes become geared not only to the external world but to each other. The world to which they became fitted included the other molecules in the body and the chemical processes in which they partook. Thereafter, when remote descendants of these marine animals moved out on to the land and became gradually more and more fitted to a dry, airy world, the old mutual adaptation of biochemical processes to each other – and incidentally to the chemical 'memory' of the sea – persisted. Why should it not, when the different kinds of molecules in the cells and blood so greatly outnumber the different kinds of molecules encountered in the outside world? It is only in a very indirect sense that the genes spell out descriptions of ancestral environments. What they directly describe, after being translated into the parallel language of protein molecules, is instructions for individual embryonic development. It is the gene pool of the species as a whole that becomes carved to fit the environments that its ancestors have encountered – which is why I said that the species is a statistical averaging device. It is in this indirect sense that our DNA is a coded description of the worlds in which our ancestors survived. And isn't it an arresting thought? We are digital archives of the African Pliocene, even of Devonian seas; walking repositories of wisdom out of the old days. You could spend a lifetime reading in this ancient library and die unsated by the wonder of it.

11

REWEAVING THE WORLD

Since my education began I have always had things described to me with their colors and sounds, by one with keen senses and a fine feeling for the significant. Therefore, I habitually think of things as colored and resonant. Habit accounts for part. The soul sense accounts for another part. The brain with its five-sensed construction asserts its right and accounts for the rest. Inclusive of all, the unity of the world demands that color be kept in it whether I have cognizance of it or not. Rather than be shut out, I take part in it by discussing it, happy in the happiness of those near to me who gaze at the lovely hues of the sunset or the rainbow.

HELEN KELLER, *The Story of My Life* (1902)

Where the gene pool of a species is sculpted into a set of models of ancestral worlds, the brain of an individual houses a parallel set of models of the animal's own world. Both are equivalent to descriptions of the past, and both are used to aid survival into the future. The difference is one of timescale and of relative privacy. The genetic description is a collective memory belonging to the species as a whole, going back into the indefinite past. The memory of the brain is private and contains the individual's experiences since it was born.

Our subjective knowledge of a familiar place does indeed feel to us like a model of the place. Not an accurate scale model, certainly less accurate than we think it is, but a serviceable model for the purposes required. One way to approach this idea was proposed some years ago by the Cambridge physiologist Horace Barlow, incidentally a direct descendant of Charles Darwin. Barlow is especially interested in vision and his argument starts from the realization that to recognize an object is a much more difficult problem than we, who seem to see so effortlessly, ordinarily understand.

For we are blissfully unaware of what a formidably clever thing we do every second of our waking lives when we see and recognize objects. The sense organs' task of unweaving the physical stimuli that bombard them is easy compared with the brain's task of reweaving an internal model of the world that it can then make use of. The argument holds for all our sensory systems, but I'll stick mostly to vision because that is the one that means the most to us.

Think what a problem our brain solves when it recognizes something, say a letter A. Or think of the problem of recognizing a particular person's face. By long in-group convention, the hypothetical face we are talking about is assumed to belong to the grandmother of the distinguished neurobiologist J. Lettvin, but substitute any face you know, or indeed any object you can recognize. We are not concerned here with subjective consciousness, with the philosophically hard problem of what it means to be *aware* of your grandmother's face. Just a cell in the brain which fires if and only if the grandmother's face appears on the retina will do nicely for a start, and it is very difficult to arrange. It would be easy if we could assume that the face would always fall exactly on a particular part of the retina. There could be a keyhole arrangement, with a grandmother-shaped region of cells on the retina wired up to a grandmother-signalling cell in the brain. Other cells – members of the 'anti-keyhole' – would have to be wired up in inhibitory fashion, otherwise the central nervous cell would respond to a white sheet just as strongly as to the grandmother's face which – together with all other conceivable images – it would necessarily 'contain'. The essence of responding to a key image is to avoid responding to everything else.

The keyhole strategy is ruled out by sheer force of numbers.

Even if Lettvin needed to recognize nothing but his grandmother, how could he cope when her image falls on a different part of the retina? How cope with her image's changing size and shape as she approaches or recedes, as she turns sideways, or cants to the rear, as she smiles or as she frowns? If we add up all possible combinations of keyholes and anti-keyholes, the number enters the astronomical range. When you realize that Lettvin can recognize not only his grandmother's face but hundreds of other faces, the other bits of his grandmother and of other people, all the letters of the alphabet, all the thousands of objects to which a normal person can instantly give a name, in all possible orientations and apparent sizes, the explosion of triggering cells gets rapidly out of hand. The American psychologist Fred Attneave, who had come up with the same general idea as Barlow, dramatized the point by the following calculation: if there were just one brain cell to cope, keyhole fashion, with each image that we can distinguish in all its presentations, the volume of the brain would have to be measured in cubic light years.

How then, with a brain capacity measured only in hundreds of cubic centimetres, do we do it? The answer was proposed in the 1950s by Barlow and Attneave independently. They suggested that nervous systems exploit the massive redundancy in all sensory information. Redundancy is jargon from the world of information theory, originally developed by engineers concerned with the economics of telephone line capacity. Information, in the technical sense, is surprise value, measured as the inverse of expected probability. Redundancy is the opposite of information, a measure of unsurprisingness, of old-hatitude. Redundant messages or parts of messages are not informative because the receiver, in some sense, already knows what is coming. Newspapers do not carry headlines saying, 'The sun rose this morning'. That would convey almost zero information. But if a morning came when the sun did not rise, headline writers would, if any survived, make much of it. The information content would be high, measured as the surprise value of the message. Much of spoken and written language is redundant –

hence possible condense telegraphese: redundancy lost, information preserved.

Everything that we know about the world outside our skulls comes to us via nerve cells whose impulses chatter like machine guns. What passes along a nerve cell is a volleying of 'spikes', impulses whose voltage is fixed (or at least irrelevant) but whose rate of arriving varies meaningfully. Now let's think about coding principles. How would you translate information from the outside world, say, the sound of an oboe or the temperature of a bath, into a pulse code? A first thought is a simple rate code: the hotter the bath, the faster the machine gun should fire. The brain, in other words, would have a thermometer calibrated in pulse rates. Actually, this is not a good code because it is uneconomical with pulses. By exploiting redundancy, it is possible to devise codes that convey the same information at a cost of fewer pulses. Temperatures in the world mostly stay the same for long periods at a time. To signal 'It is hot, it is hot, it is still hot . . .' by a continuously high rate of machine-gun pulses is wasteful; it is better to say, 'It has suddenly become hot' (now you can assume that it will stay the same until further notice).

And, satisfyingly, this is what nerve cells mostly do, not just for signalling temperature but for signalling almost everything about the world. Most nerve cells are biased to signal *changes* in the world. If a trumpet plays a long sustained note, a typical nerve cell telling the brain about it would show the following pattern of impulses: Before the trumpet starts, low firing rate; immediately after the trumpet starts, high firing rate; as the trumpet carries on sustaining its note, the firing rate dies away to an infrequent mutter; at the moment when the trumpet stops, high firing rate, dying away to a resting mutter again. Or there might be one class of nerve cells that fire only at the onset of sounds and a different class of cells that fire only when sounds go off. Similar exploitation of redundancy – screening out of the *sameness* in the world – goes on in cells that tell the brain about changes in light, changes in temperature, changes in pressure. Everything about the world is signalled as *change*, and this is a major economy.

But you and I don't seem to hear the trumpet die away. To us the trumpet seems to carry on at the same volume and then to stop abruptly. Yes, of course. That's what you'd expect because the coding system is ingenious. It doesn't throw away information, it only throws away redundancy. The brain is told only about changes, and it is then in a position to reconstruct the rest. Barlow doesn't put it like this, but we could say that the brain constructs a virtual sound, using the messages supplied by the nerves coming from the ears. The reconstructed virtual sound is complete and unabridged, even though the messages themselves are economically stripped down to information about changes. The system works because the state of the world at a given time is usually not greatly different from the preceding second. Only if the world changed capriciously, randomly and frequently, would it be economical for sense organs to signal continuously the state of the world. As it is, sense organs are set up to signal, economically, the discontinuities in the world; and the brain, assuming correctly that the world doesn't change capriciously and at random, uses the information to construct an internal virtual reality in which the continuity is restored.

The world presents an equivalent kind of redundancy in space, and the nervous system uses the corresponding trick. Sense organs tell the brain about edges and the brain fills in the boring bits between. Suppose you are looking at a black rectangle on a white background. The whole scene is projected on to your retina – you can think of the retina as a screen covered with a dense carpet of tiny photocells, the rods and cones. In theory, each photocell could report to the brain the exact state of the light falling upon it. But the scene we are looking at is massively redundant. Cells registering black are overwhelmingly likely to be surrounded by other cells registering black. Cells registering white are nearly all surrounded by other white-signalling cells. The important exceptions are cells on edges. Those on the white side of an edge signal white themselves and so do their neighbours that sit further into the white area. But their neighbours on the other side are in the black area. The brain can theoretically reconstruct the whole scene if just the retinal cells on edges fire. If this could be achieved there would be massive savings in nerve impulses. Once again, redundancy is removed and only information gets through.

Elegantly, the economy is achieved in practice by the mechanism known as 'lateral inhibition'. Here's a simplified version of the principle, using our analogy of the screen of photocells. Each photocell sends one long wire to the central computer (brain) and also short wires to its immediate neighbours in the photocell screen. The short connections to the neighbours inhibit them, that is, turn down their firing rate. It is easy to see that maximal firing will come only from cells that lie along edges, for they are inhibited from one side only. Lateral inhibition of this kind is common among the low-level units of both vertebrate and invertebrate eyes.

Once again, we could say that the brain constructs a virtual world which is more complete than the picture relayed to it by the senses. The information which the senses supply to the brain is mostly information about edges. But the model in the brain is able to reconstruct the bits between the edges. As in the case of discontinuities in time, an economy is achieved by the elimination – and later reconstruction in the brain – of redundancy. This economy is possible only because uniform patches exist in the world. If the shades and colours in the world were randomly dotted about, no economical remodelling would be possible.

Another kind of redundancy stems from the fact that many lines in the real world are straight, or curved in smooth and therefore predictable (or mathematically reconstructable), ways. If the ends of a line are specified, the middle can be filled in using a simple rule that the brain already 'knows'. Among the nerve cells that have been discovered in the brains of mammals are the so-called 'line-detectors', neurones that fire whenever a straight line, aligned in a particular direction, falls on a particular place in the retina, the so-called 'retinal field' of the brain cell. Each of these line-detector cells has its own preferred direction. In the cat brain, there are only two preferred directions, horizontal and vertical, with an approximately equal number favouring each direction; however, in monkeys other angles are accommodated. From the point of view of the redundancy argument, what is going on here is as follows. In the retina, all the cells along a straight line fire and most of these impulses are redundant. The nervous system economizes by using a single cell to register the line, labelled with its angle. Straight lines are economically specified by their position and direction alone, or by their ends, not by the light value of every point along their length. The brain reweaves a virtual line in which the points along the line are reconstructed.

However, if a part of a scene suddenly detaches itself from the rest and starts to crawl over the background, it is news and should be signalled. Biologists have indeed discovered nerve cells that are silent until something moves against a still background. These cells don't respond when the entire scene moves – that would correspond to the sort of apparent movement the animal would see when it itself moves. But movement of a small object against a still background is information-rich and there are nerve cells tuned to detect it. The most famous of these are the so-called 'bug-detectors' discovered in frogs by Lettvin (he of the grandmother) and his colleagues. A bug-detector is a cell which is apparently blind to everything except the movement of small objects against their background. As soon as an insect moves in the field covered by a bug-detector, the cell immediately initiates massive signalling and the frog's tongue is likely to shoot out to catch the insect. To a sufficiently sophisticated nervous system, though, even the movement of a bug is redundant if it is movement in a straight line. Once you've been told that a bug is moving steadily in a northerly direction, you can assume that it will continue to move in this direction until further notice.

Carrying the logic a step further, we should expect to find higher-order movement detector cells in the brain that are especially sensitive to *change* in movement, say, change in direction or change in speed. Lettvin and his colleagues found a cell that seems to do this, again in the frog. In their paper in *Sensory Communication* (1961) they describe a particular experiment as follows:

Let us begin with an empty gray hemisphere for the visual field. There is usually no response of the cell to turning on and off the illumination. It is silent. We bring in a small dark object, say 1 to 2 degrees in diameter, and at a certain point in its travel, almost anywhere in the field, the cell suddenly 'notices' it. Thereafter, wherever that object is moved it is tracked by the cell. Every time it moves, with even the faintest jerk, there is a burst of impulses that dies down to a mutter that continues as long as the object is visible. If the object is kept moving, the bursts signal discontinuities in the movement, such as the turning of corners, reversals, and so forth, and these bursts occur against a continuous background mutter that tells us the object is visible to the cell . . .

To summarize, it is as if the nervous system is tuned at successive hierarchical levels to respond strongly to the unexpected, weakly or not at all to the expected. What happens at successively higher levels is that the definition of that which is expected becomes progressively more sophisticated. At the lowest level, every spot of light is news. And the next level up, only edges are 'news'. At a higher level still, since so many edges are straight, only the ends of edges are news. Higher again, only movement is news. Then only changes in rate or direction of movement. In Barlow's terms derived from the theory of codes, we could say that the nervous system uses short, economical words for messages that occur frequently and are expected; long, expensive words for messages that occur rarely and are not expected. It is a bit like language, in which (the generalization is called Zipf's Law) the shortest words in the dictionary are the ones most often used in speech. To push the idea to an extreme, most of the time the brain does not need to be told anything because what is going on is the norm. The message would be redundant. The brain is protected from redundancy by a hierarchy of filters, each filter tuned to remove expected features of a certain kind.

It follows that the set of nervous filters constitutes a kind of summary description of the norm, of the statistical properties of the world in which the animal lives. It is the nervous equivalent of our insight of the previous chapter: that the genes of a species come to constitute a statistical description of the worlds in which its ancestors were naturally selected. Now we see that the sensory coding units with which the brain confronts the environment also constitute a statistical description of that environment. They are tuned to discount the common and emphasize the rare. Our hypothetical zoologist of the future should therefore be able, by inspecting the nervous system of an unknown animal and measuring the statistical biases in its tuning, to reconstruct the statistical properties of the world in which the animal lived, to read off what is common and what rare in the animal's world.

The inference would be indirect, in the same way as for the case of the genes. We would not be reading the animal's world as a direct description. Rather, we'd infer things about the animal's world by inspecting the glossary of abbreviations that its brain used to describe it. Civil servants love acronyms like CAP (Common Agricultural Policy) and HEFCE (Higher Education Funding Council for England); fledgling bureaucrats surely need a glossary of such abbreviations, a codebook. If you find such a codebook dropped in the street, you could work out which ministry it came from by seeing which phrases have been granted abbreviations, presumably because they are commonly used in that ministry. An intercepted codebook is not a particular message about the world, but it is a statistical summary of the kind of world which this code was designed to describe economically.

We can think of each brain as equipped with a store cupboard of basic images, useful for modelling important or common features of the animal's world. Although, following Barlow, I have emphasized learning as the means by which the store cupboard is stocked, there is no reason why natural selection itself, working on genes, should not do some of the work of filling up the cupboard. In this case, following the logic of the previous chapter, we should say that the store cupboard in the brain contains images from the ancestral past of the species. We could call it a collective unconscious, if the phrase had not become tarnished by association.

But the biases of the image kit in the cupboard will not only reflect what is statistically unexpected in the world. Natural selection will ensure that the repertoire of virtual representations is also well endowed with images that are of particular salience or importance in the life of the particular kind of animal and in the world of its ancestors, even if these are not especially common. An animal may need only once in its life to recognize a complicated pattern, say the shape of a female of its species, but on that occasion it is vitally important to get it right, and do so without delay. For humans, faces are of special importance, as well as being common in our world. The same is true of social monkeys. Monkey brains have been found to possess a special class of cells which fire at full strength only when presented with a complete face. We've already seen that humans with particular kinds of localized brain damage experience a very peculiar, and revealing, kind of selective blindness. They can't recognize faces. They can see everything else, apparently normally, and they can see that a face has a shape, with features. They can describe the nose, the eyes and the mouth. But they can't recognize the face even of the person they love best in all the world.

Normal people not only recognize faces. We seem to have an almost indecent eagerness to see faces, whether they are really there or not. We see faces in damp patches on the ceiling, in the contours of a hillside, in clouds or in Martian rocks. Generations of moongazers have been led, by the most unpromising of raw materials, to invent a face in the pattern of craters on the moon. The *Daily Express* (London) of 15 January 1998 bestowed most of a page, complete with banner headline, on the story that an Irish cleaning woman saw the face of Jesus in her duster: 'Now a stream of pilgrims is expected at her semi-detached home . . . The woman's parish priest said, "I've never seen anything like it before in my 34 years in the priesthood."' The accompanying photograph shows a pattern of dirty polish on a cloth which slightly resembles a face of some kind: there is a faint suggestion of an eye on one side of what could be a nose; there is also a sloping eyebrow on the other side which gives it a look of Harold Macmillan, although I suppose even Harold Macmillan might look like Jesus to a suitably prepared mind. The *Express* reminds us of similar stories, including the 'nun bun' served up in a Nashville café, which 'resembled the face of Mother Teresa, 86' and caused great excitement until 'the aged nun wrote to the café demanding the bun be removed'.

The eagerness of the brain to construct a face, when offered the slightest encouragement, fosters a remarkable illusion. Get an ordinary mask of a human face – President Clinton's face, or whatever is on sale for fancy dress parties. Stand it up in a good light and look at it from the far side of the room. If you look at it the normal way round, not surprisingly it looks solid. But now turn the mask so that it is facing away from you and look at the hollow side from across the room. Most people see the illusion immediately. If you don't, try adjusting the light. It may help if you shut one eye, but it is by no means necessary. The illusion is that the hollow side of the mask looks solid. The nose, brows and mouth stick out towards you and seem nearer than the ears. It is even more striking if you move from side to side, or up and down. The apparently solid face seems to turn with you, in an odd, almost magical way.

I'm not talking about the ordinary experience we have when the eyes of a good portrait seem to follow you around the room. The hollow mask illusion is far more spooky. It seems to hover, luminously, in space. The face really *really* seems to turn. I have a mask of Einstein's face mounted in my room, hollow side out, and visitors gasp when they glimpse it. The illusion is most strikingly displayed if you set the mask on a slowly rotating turntable. As the solid side turns before you, you'll see it move in a sensible 'normal reality' way. Now the hollow side comes into view and something extraordinary happens. You see another solid face, but it is rotating in the opposite direction. Because one face (say, the real solid face) is turning clockwise while the other, pseudo-solid face appears to be turning anticlockwise, the face that is rotating into view seems to swallow up the face that is rotating away from view. As the turning continues, you then see the really hollow but apparently solid face rotating firmly in the wrong direction for a while, before the really solid face reappears and swallows up the virtual face. The whole experience of watching the illusion is quite unsettling and it remains so no matter how long you go on watching it. You don't get used to it and don't lose the illusion.

What is happening? We can take the answer in two stages. First, why do we see the hollow mask as solid? And second, why does it seem to rotate in the wrong direction? We've already agreed that the brain is very good at – and very keen on – constructing faces in its internal simulation room. The information that the eyes are feeding to the brain is of course compatible with the mask's being hollow, but it is also compatible – just – with an alternative hypothesis, that it is solid. And the brain, in its simulation, goes for the second alternative, presumably because of its eagerness to see faces. So it overrules the messages from the eyes that say, 'This is hollow'; instead, it listens to the messages that say, 'This is a face, this is a face, face, face, face.' Faces are always solid. So the brain takes a face model out of its cupboard which is, by its nature, solid.

But having constructed its apparently solid face model, the brain is caught in a contradiction when the mask starts to rotate. To simplify the explanation, suppose that the mask is that of Oliver Cromwell and that his famous warts are visible from both sides of the mask. When looking at the hollow interior of the nose, which is really pointing away from the viewer, the eye looks straight across to the right side of the nose where there is a prominent wart. But the constructed virtual nose is apparently pointing towards the viewer, not away, and the wart is on what, from the virtual Cromwell's point of view, would be his left side, as if we were looking at Cromwell's mirror image. As the mask rotates, if the face were really solid, our eye would see more of the side that it expected to see more of and less of the side that it expected to see less of. But because the mask is actually hollow, the reverse happens. The relative proportions of the retinal image change in the way the brain would expect if the face were solid but rotating in the opposite direction. And that is the illusion that we see. The brain resolves the inevitable contradiction, as one side gives way to the other, in the only way possible, given its stubborn insistence on the mask's being a solid face: it simulates a virtual model of one face swallowing up the other face.

The rare brain disorder that destroys our ability to recognize faces is called prosopagnosia. It is caused by injury to specific parts of the brain. This very fact supports the importance of a 'face cupboard' in the brain. I don't know, but I'd bet that prosopagnosics wouldn't see the hollow mask illusion. Francis Crick discusses prosopagnosia in his book *The Astonishing Hypothesis* (1994), together with other revealing clinical conditions. For instance, one patient found the following condition very frightening which, as Crick observes, is not surprising:

> *... objects or persons she saw in one place suddenly appeared in another without her being aware they were moving. This was particularly distressing if she wanted to cross a road, since a car that at first seemed far away would suddenly be very close ... She experienced the world rather as some of us might see the dance floor in the strobe lighting of a discotheque.*

This woman had a mental cupboard full of images for assembling her virtual world, just as we all do. The images themselves were probably perfectly good. But something had gone wrong with her software for deploying them in a smoothly changing virtual world. Other patients have lost their ability to construct virtual depth. They see the world as though it was made of flat, cardboard cut-outs. Yet other patients can recognize objects only if they are presented from a familiar angle. The rest of us, having seen, say, a saucepan from the side, can effortlessly recognize it from above. These patients have presumably lost some ability to manipulate virtual images and turn them around. The technology of virtual reality gives us a language to think about such skills, and this will be my next topic.

I shall not dwell on the details of today's virtual reality which is certain, in any case, to become obsolete. The technology changes as rapidly as everything else in the world of computers. Essentially what happens is as follows. You don a headset which presents to each of your eyes a miniature computer screen. The images on the two screens are nearly the same as each other, but offset to give the stereo illusion of three dimensions. The scene is whatever has been programmed into the computer: the Parthenon, perhaps, intact and in its original garish colours; an imagined landscape on Mars; the inside of a cell, hugely magnified. So far, I might have been describing an ordinary 3-D movie. But the virtual reality machine provides a two-way street. The computer doesn't just present you with scenes, it responds to you. The headset is wired up to register all turnings of your head, and other body movements, which would, in the normal course of events, affect your viewpoint. The computer is continuously informed of all such movements and – here is the cunning part – it is programmed to change the scene presented to the eyes, in exactly the way it would change if you were really moving your head. As you turn your head, the pillars of the Parthenon, say, swing round and you find yourself looking at a statue which, previously, had been 'behind' you.

A more advanced system might have you in a body stocking, laced with strain gauges to monitor the positions of all your limbs. The computer can now tell whenever you take a step, whenever you sit down, stand up, or wave your arms. You can now walk from one end of the Parthenon to the other, watching the pillars pass by as the computer changes the images in sympathy with your steps. Tread carefully because, remember, you are not really in the Parthenon but in a cluttered computer room. Present day virtual reality systems, indeed, are likely to tether you to the computer by a complicated umbilicus of cables, so let's postulate a future tangle-free radio link, or infrared data beam. Now you can walk freely in an empty real world and explore the fantasy virtual world that has been programmed for you. Since the computer knows where your body stocking is, there is no reason why it shouldn't represent you to yourself as a complete human form, an *avatar*, allowing you to look down at your 'legs', which might be very different from your real legs. You can watch your avatar's hands as they move in imitation of your real hands. If you use these hands to pick up a virtual object, say a Grecian urn, the urn will seem to rise into the air as you 'lift' it.

If somebody else, who could be in another country, dons another set of kit hooked up to the same computer, in principle you should be able to see their avatar and even shake hands – though with present day technology you might find yourself passing through each other like ghosts. The technicians and programmers are still working on how to create the illusion of texture and the 'feel' of solid resistance. When I visited England's leading virtual reality company, they told me they get many letters from people wanting a virtual sexual partner. Perhaps in the future, lovers separated by the Atlantic will caress each other over the Internet, albeit incommoded by the need to wear gloves and a body stocking wired up with strain gauges and pressure pads.

Now let's take virtual reality a shade away from dreams and closer to practical usefulness. Present day doctors have recourse to the ingenious endoscope, a sophisticated tube that is inserted into a patient's body through, say, the mouth or the rectum and used for diagnosis and even surgical intervention. By the equivalent of pulling wires, the surgeon steers the long tube round the bends of the intestine. The tube itself has a tiny television camera lens at its tip and a light pipe to illuminate the way. The tip of the tube may also be furnished with various remote-control instruments which the surgeon can control, such as micro-scalpels and forceps.

In conventional endoscopy, the surgeon sees what he is doing using an ordinary television screen, and he operates the remote controls using his fingers. But as various people have realized (not least Jaron Lanier, who coined the phrase 'virtual reality' itself) it is in principle possible to give the surgeon the illusion of being shrunk and actually inside the patient's body. This idea is in the research stage, so I shall resort to a fantasy of how the technique might work in the next century. The surgeon of the future has no need to scrub up, for she need not go near her patient. She stands in a wide open area, connected by radio to the endoscope inside the patient's intestine. The miniature screens in front of her two eyes present a magnified stereo image of the interior of the patient immediately in front of the tip of the endoscope. When she moves her head to the left, the computer automatically swivels the tip of the endoscope to the left. The angle of view of the camera inside the intestine faithfully moves to follow the surgeon's head movements in all three planes. She drives the endoscope forward along the intestine by her footsteps. Slowly, slowly, for fear of damaging the patient, the computer pushes the endoscope forwards, its direction always controlled by the direction in which, in a completely different room, the surgeon is walking. It feels to her as though she is actually walking through the intestine. It doesn't even feel claustrophobic. Following present day endoscopic practice, the gut has been carefully inflated with air, otherwise the walls would press in upon the surgeon and force her to crawl rather than walk.

When she finds what she is looking for, say a malignant tumour, the surgeon selects an instrument from her virtual toolbag. Perhaps it is most convenient to model it as a chainsaw, whose image is generated in the computer. Looking through the stereo screens in her helmet at the enlarged 3-D tumour, the surgeon sees the virtual chainsaw in her virtual hands and goes to work, excising the tumour, as though it were a tree stump needing to be removed from the garden. Inside the real patient, the mirrored equivalent of the chainsaw is an ultrafine laser beam. As if by a pantograph, the gross movements of the surgeon's whole arm as she hefts the chainsaw are geared down, by the computer, to equivalent tiny movements of the laser gun in the tip of the endoscope.

For my purposes I need say only that it is theoretically possible to create the illusion of walking through somebody's intestine using the techniques of virtual reality. I do not know whether it will actually help surgeons. I suspect that it will, although a present day hospital consultant whom I have asked is a little sceptical. This same surgeon refers to himself and his fellow gastroenterologists as glorified plumbers. Plumbers themselves sometimes use larger-scale versions of endoscopes for exploring pipes and in America they even send down mechanical 'pigs' to eat their way through blockages in drains. Obviously the methods I imagined for a surgeon would work for a plumber. The plumber could 'tramp' (or 'swim'?) down the virtual water pipe with a virtual miner's lamp on his helmet and a virtual pickaxe in his hand for clearing blockages.

The Parthenon of my first example existed nowhere but in the computer. The computer could as well have introduced you to angels, harpies or winged unicorns. My hypothetical endoscopist and plumber, on the other hand, were walking through a virtual world that was constrained to resemble a mapped portion of reality, the real interior of a drain or a patient's intestine. The virtual world that was presented to the surgeon on her stereo screens was admittedly constructed in a computer, but it was constructed in a disciplined way. There was a real laser gun being controlled, albeit represented as a chainsaw because this would feel like a natural tool to excise a tumour whose apparent size was comparable to the surgeon's own body. The shape of the virtual construction reflected, in the way most convenient to the surgeon's operation, a detail of the real world inside the patient. Such *constrained* virtual reality is pivotal in this chapter. I believe that every species that has a nervous system uses it to construct a model of its own particular world, constrained by continuous updating through the sense organs. The nature of the model may depend upon how the species concerned is going to use it, at least as much as upon what we might think of as the nature of the world itself.

Think of a gliding gull adroitly riding the winds off a sea cliff. It may not be flapping its wings, but this doesn't mean that its wing muscles are idle. They and the tail muscles are constantly making tiny adjustments, sensitively fine-tuning the bird's flight surfaces to every eddy, every nuance of the air around it. If we fed information about the state of all the nerves controlling these muscles into a computer, from moment to moment, the computer could in principle reconstruct every detail of the air currents through which the bird was gliding. It would do this by assuming that the bird was well designed to stay aloft and on that assumption construct a continuously updated model of the air around it. It would be a dynamic model, like a weather forecaster's model of the world's weather system, which is continuously revised by new data supplied by weather ships, satellites and ground stations and can be extrapolated to predict the future. The weather model advises us about tomorrow's weather; the gull model is theoretically capable of 'advising' the bird on the anticipatory adjustments that it should make to its wing and tail muscles in order to glide on into the next second.

The point we are working towards, of course, is that although no human programmer has yet constructed a computer model to advise gulls on how to adjust their wing and tail muscles, just such a model is surely being run continuously in the brain of our gull and of every other bird in flight. Similar models, pre-programmed in outline by genes and past experience, and continuously updated by new sense data from millisecond to millisecond, are running inside the skull of every swimming fish, every galloping horse, every echo-ranging bat.

That ingenious inventor Paul MacCready is best known for his superbly economical flying machines, the man-powered *Gossamer Condor* and *Gossamer Albatross*, and the sun-powered *Solar Challenger*. He also, in 1985, constructed a half-sized flying replica of the giant Cretaceous pterosaur *Quetzalcoatlus*. This huge flying reptile, with a wingspan comparable to that of a light aircraft, had almost no tail and was therefore highly unstable in the air. John Maynard Smith, who trained as an aero-engineer before switching to zoology, pointed out that this would have given advantages of manoeuvrability, but it demands accurate moment-to-moment control of the flight surfaces. Without a fast computer to adjust its trim continuously, MacCready's replica would have crashed. The real *Quetzalcoatlus* must have had an equivalent computer in its head, for the same reason. Earlier pterosaurs had long tails, in some cases terminated by what looks like a ping-pong bat, which would have given great stability, at a cost in manoeuvrability. It seems that, in the evolution of late, almost tailless pterosaurs like *Quetzalcoatlus*, there was a shift from stable but unmanoeuvrable to manoeuvrable but unstable. The same trend can be seen in the evolution of manmade aeroplanes. In both cases, the trend is made possible only by increasing computer power. As in the case of the seagull, the pterosaur's on-board computer inside its skull must have run a simulation model of the animal and the air through which it flew.

You and I, we humans, we mammals, we animals, inhabit a virtual world, constructed from elements that are, at successively higher levels, useful for representing the real world. Of course, we feel as if we are firmly placed in the real world – which is exactly as it should be if our constrained virtual reality software is any good. It is very good, and the only time we notice it at all is on the rare occasions when it gets something wrong. When this happens we experience an illusion or a hallucination, like the hollow mask illusion we talked about earlier.

The British psychologist Richard Gregory has paid special attention to visual illusions as a means of studying how the brain works. In his book *Eye and Brain* (fifth edition 1998), he regards seeing as an active process in which the brain sets up hypotheses about what is going on out there, then tests those hypotheses against the data coming in from the sense organs. One of the most familiar of all visual illusions is the Necker cube. This is a simple line drawing of a hollow cube, like a cube made of steel rods. The drawing is a two-dimensional pattern of ink on paper. Yet a normal human sees it as a cube. The brain has made a three-dimensional model based upon the two-dimensional pattern on the paper. This is, indeed, the kind of thing the brain does almost every time you look at a picture. The flat pattern of ink on paper is equally compatible with two alternative three-dimensional brain models. Stare at the drawing for some seconds and you will see it flip. The facet that had previously seemed nearest to you will now appear farthest. Carry on looking, and it will flip back to the original cube. The brain could have been designed to stick, arbitrarily, to one of the two cube models, say the first of the two that it hit upon, even though the other model would have been equally compatible with the information from the retinas. But in fact the brain takes the other option of running each model, or hypothesis, alternately for a few seconds at a time. Hence the apparent cube alternates, which gives the game away. Our brain constructs a three-dimensional model. It is virtual reality in the head.

When we are looking at an actual wooden box, our simulation software is provided with additional information, which enables it to arrive at a clear preference for one of the two internal models. We therefore see the box in one way only, and there is no alternation. But this does not diminish the truth of the general lesson we learn from the Necker cube. Whenever we look at anything, there is a sense in which what our brain actually makes use of is a model of that thing in the brain. The model in the brain, like the virtual Parthenon of my earlier example, is constructed. But, unlike the Parthenon (and perhaps the visions we see in dreams), it is, like the surgeon's computer model of the inside of her patient, not entirely invented: it is constrained by information fed in from the outside world.

A more powerful illusion of solidity is conveyed by stereoscopy, the slight discrepancy between the two images seen by the left and the right eyes. It is this that is exploited by the two screens in a virtual reality helmet. Hold up your right hand, with the thumb towards you, about one foot in front of your face, and look at some distant object, say a tree, with both eyes open. You'll see two hands. These correspond to the images seen by your two eyes. You can quickly find out which is which by first shutting one, then the other, eye. The two hands appear to be in slightly different places because your two eyes converge from different angles and the images on the two retinas are correspondingly, and tellingly, different. The two eyes get a slightly different view of the hand, too. The left eye sees a bit more of the palm, the right eye sees a bit more of the back of the hand.

Now, instead of looking at the distant tree, look at your hand, again with both eyes open. Instead of two hands in the foreground and one tree in the background, you'll see one solid-looking hand and two trees. Yet the hand image is still falling on different places on your two retinas. What this means is that your simulation software has *constructed* a single model of the hand, a model in 3-D. What's more, the single three-dimensional model has made use of information from both eyes. The brain subtly amalgamates both sets of information and puts together a useful model of a single, three-dimensional, solid hand. Incidentally, all retinal images of course are upside down, but this doesn't matter because the brain constructs its simulation model in the way that best suits its purpose and defines this model as the right way up.

The computational tricks used by the brain to construct a three-dimensional model from two two-dimensional images are astonishingly sophisticated, and are the basis of perhaps the most impressive of all illusions. These date back to a discovery by the Hungarian psychologist Bela Julesz in 1959. A normal stereoscope presents the same photograph to the left and the right eye but taken from suitably different angles. The brain puts the two together and sees an impressively three-dimensional scene. Julesz did the same thing, except that his pictures were random pepper and salt dots. The left and the right eye were shown the same random pattern, but with a crucial difference. In a typical Julesz experiment, an area of the pattern, say, a square, has its random dots displaced to one side, the appropriate distance to create the stereoscopic illusion. And the brain sees the illusion – a square patch stands out – even though there is not the smallest trace of a square in either of the two pictures. The square is present *only* in the discrepancy between the two pictures. The square looks very real to the viewer, but it really is nowhere but in the brain. The Julesz Effect is the basis of the 'Magic Eye' illusions so popular today. In a *tour de force* of the explainer's art, Steven Pinker devotes a small section of *How the Mind Works* (1998) to the principle underlying these pictures. I won't even try to better his explanation.

There is an easy way to demonstrate that the brain works as a sophisticated virtual reality computer. First, look about you by moving your eyes. As you swivel your eyes, the images on your retinas move as if you were in an earthquake. But you don't see an earthquake. To you, the scene seems as steady as a rock. I am leading up, of course, to saying that the virtual model in your brain is constructed to remain steady. But there is more to the demonstration, because there's another way to make the image on your retina move. Gently poke your eyeball through the skin of the eyelid. The retinal image will move in the same kind of way as before. Indeed you could, given sufficient skill with your finger, mimic the effect of shifting your gaze. But now you really will think you see the earth move. The whole scene shifts, as if you were witnessing an earthquake.

What is the difference between these two cases? It is that the brain computer has been set up to take account of normal eye movements and make allowance for them in constructing its computed model of the world. Apparently the brain model makes use of information, not only from the eyes, but also from the instructions to move the eyes. Whenever the brain issues an order to the eye muscles to move the eye, a copy of that order is sent to the part of the brain that is constructing the internal model of the world. Then, when the eyes move, the virtual reality software of the brain is warned to *expect* the retinal images to move just the right amount, and it makes the model compensate. So the constructed model of the world is seen to stay still, although it may be viewed from another angle. If the earth moves at any time other than when the model is told to expect movement, the virtual model moves accordingly. This is fine, because there really might be an earthquake. Except that you can fool the system by poking your eyeball.

As the final demonstration using yourself as guinea pig, make yourself giddy by spinning round and round. Now stand still and look fixedly at the world. It will appear to spin even though your reason tells you that it is not getting anywhere in its rotation. Your retinal images are not moving, but the accelerometers in your ears (which work by detecting the movements of fluid in the so-called semicircular canals) are telling the brain that you are spinning. The brain instructs the virtual reality software to expect to see the world spinning. When the images on the retina do not spin, therefore, the model registers the discrepancy and spins itself in the opposite direction. To put it in subjective language, the virtual reality software says to itself, 'I know I'm spinning from what the ears are telling me; therefore, in order to hold the model still, it will be necessary to put the opposite spin on the model, relative to the data that the eyes are sending in.' But the retinas actually report no spin, so the compensating spin of the model in the head is what you seem to see. In Barlow's terms, it is the unexpected, it is 'news', and that is why we see it.

Birds have an additional problem which humans ordinarily are spared. A bird perched on a tree branch is constantly being blown up and down, to and fro, and its retinal images seesaw accordingly. It is like living through a permanent earthquake. Birds keep their heads, and hence their view of the world, steady by diligent use of the neck muscles. If you film a bird on a windblown branch, you can almost imagine that the head is nailed to the background, while the neck muscles use the head as a fulcrum to move the rest of the body. When a bird walks, it employs the same trick to keep its perceived world steady. That is why walking chickens jerk their heads back and forth in what can seem to us quite a comical fashion. It is actually rather clever. As the body moves forward, the neck draws the head backwards in a controlled way so that the retinal images remain steady. Then the head shoots forward to allow the cycle to repeat. I can't help wondering whether, as an untoward consequence of the bird way of doing things, a bird might be unable to see a real earthquake because its neck muscles would automatically compensate. More seriously, we might say that the bird is using its neck muscles in a Barlow-style exercise: holding the non-newsworthy part of the world constant so that genuine movement stands out.

Insects and many other animals seem to have a similar habit of working to keep their visual world constant. Experimenters have demonstrated this in a so-called 'optomotor apparatus', where the insect is placed on a table and surrounded by a hollow cylinder painted on the inside with vertical stripes. If you now rotate the cylinder, the insect will use its legs to turn, keeping up with the cylinder. It is working to keep its visual world constant.

Normally, an insect has to tell its simulating software to expect movement when it walks, otherwise it would start compensating for its own movements, and then where would it be? This thought prompted two ingenious Germans, Erich von Holst and Horst

Mittelstaedt, to a diabolically cunning experiment. If you've ever watched a fly washing its face with its hands, you will know that flies are capable of flicking their head completely upside down. Von Holst and Mittelstaedt succeeded in fixing a fly's head in the inverted position using glue. You have already guessed the consequence. Normally, whenever a fly turns its body, the model in its brain is told to expect a corresponding movement of the visual world. But as soon as it took a step, the wretched fly with its head upside down received data suggesting that the world had moved in the opposite direction to the one expected. It therefore moved its legs further in the same direction in order to compensate. This caused the apparent position of the world to move even further. The fly ended up spinning round and round like a top, at ever-increasing speed – well, within obvious practical limits.

The same Erich von Holst also pointed out that we should expect a similar confusion if our own voluntary instructions to move our eyes are neutralized, for example by narcotizing the eye-moving muscles. Normally, if you give your eyes the command to move to the right, your retinal images will signal a move to the left. To compensate and create the appearance of stability, the model in the head has to be moved to the right. But if the eye-moving muscles are narcotized, the model should move to the right in anticipation of what turns out to be a non-existent retinal movement. Let von Holst himself take up the story, in his paper 'The Behavioural Physiology of Animals and Man' (1973):

This is indeed the case! It has been known for many years from people with paralysed eye muscles and it has been established exactly from the experiments of Kornmuller on himself that every intended but unfulfilled eye movement results in the perception of a quantitative movement of the surroundings in the same direction.

We are so used to living in our simulated world and it is kept so beautifully in synchrony with the real world that we don't realize it is a simulated world. It takes clever experiments like those of von Holst and his colleagues to bring it home to us.

And it has its dark side. A brain that is good at simulating models in imagination is also, almost inevitably, in danger of self-delusion. How many of us as children have lain in bed, terrified because we thought we saw a ghost or a monstrous face staring in at the bedroom window, only to discover that it was a trick of the light? I've already discussed how eagerly our brain's simulation software will construct a solid face where the reality is a hollow face. It will just as eagerly make a ghostly face where the reality is a collection of moonlit folds in a white net curtain.

Every night of our lives we dream. Our simulation software sets up worlds that do not exist; people, animals and places that never existed, perhaps never could exist. At the time, we experience these simulations as though they were reality. Why should we not, given that we habitually experience reality in the same way – as simulation models? The simulation software can delude us when we are awake, too. Illusions like the hollow face are in themselves harmless, and we understand how they work. But our simulation software can also, if we are drugged, or feverish, or fasting, produce hallucinations. Throughout history, people have seen visions of angels, saints and gods; and these have seemed very real to them. Well, of course they *would* seem real. They are models, put together by the normal simulation software. The simulation software is using the same modelling techniques as it uses ordinarily when it presents its continuously updated edition of reality. No wonder these visions have been so influential. No wonder they have changed people's lives. So if ever we hear a story that somebody has seen a vision, been visited by an archangel, or heard voices in the head, we should immediately be suspicious of taking it at face value. Remember that all our heads contain powerful and ultra-realistic simulation software. Our simulation software could knock up a ghost or a dragon or a saintly virgin in no time flat. It would be child's play for software of that sophistication.

A word of warning. The metaphor of virtual reality is beguiling and, in many ways, apt. But there is a danger of its misleading us into thinking that there is a 'little man' or 'homunculus' in the brain watching the virtual reality show. As philosophers such as Daniel Dennett have pointed out, you have explained precisely nothing if you suggest that the eye is wired to the brain in such a way that a little cinema screen, somewhere in the brain, continuously relays whatever is projected on the retina. Who looks at the screen? The question now raised is no smaller than the original question you think you have answered. You might as well let the little man look at the retina directly, which is clearly no solution to anything. The same problem arises if we take the virtual reality metaphor literally and imagine that some agent locked inside the head is 'experiencing' the virtual reality performance.

The problems raised by subjective consciousness are perhaps the most baffling in all philosophy, and solving them is far beyond my ambition. My suggestion is the more modest one that each species, in each situation, needs to deploy its information about the world in whatever way is most useful for taking action. 'Constructing a model in the head' is a helpful way to express how it is done, and comparing it to virtual reality is especially helpful in the case of humans. As I have argued before, the model of the world used by a bat is likely to be similar to the model used by a swallow, even though one is connected to the real world via the ears, the other via the eyes. The brain constructs its model world in the way most suited for action. Since the actions of day-flying swallows and night-flying bats are similar – navigating at high speed in three dimensions, avoiding solid obstacles and catching insects on the wing – they are likely to use the same models. I do not postulate a 'little bat in the head' or a 'little swallow in the head' to watch the model. Somehow the model is used to control the wing muscles, and that is as far as I go.

Nevertheless, each of us humans knows that the illusion of a single agent sitting somewhere in the middle of the brain is a powerful one. I suspect that the case may be parallel to the 'selfish cooperator' model of genes coming together, although they are fundamentally independent agents, to create the illusion of a unitary body. I'll briefly return to the idea near the end of the next chapter.

This chapter has developed the thesis that brains have taken over from DNA part of the role of recording the environment – environments, rather, for they are many and spread out over the near and the distant past. Having a record of the past is useful only in so far as it helps in predicting the future. The animal's body represents a kind of prediction that the future will resemble the ancestral past, in broad outline. The animal is likely to survive to the extent that this turns out to be true. And simulation models of the world allow the animal to act as if in anticipation of what that world is likely to throw its way in the next few seconds, hours or days. For completeness we must note that the brain itself, and its virtual reality software, are ultimately the products of natural selection of ancestral genes. We could say that the genes can predict a limited amount, because only in a general way will the future resemble the past. For the details and the subtleties, they provide the animal with nervous hardware and virtual reality software which will constantly update and revise its predictions to fit high-speed changes in circumstances. It is as if the genes say, 'We can model the basic shape of the environment, the things that don't change over the generations. But for the fast changes, over to you, brain.'

We move through a virtual world of our own brains' making. Our constructed models of rocks and of trees are a part of the environment in which we animals live, no less than the real rocks and trees that they represent. And, intriguingly, our virtual worlds must also be seen as part of the environment in which our genes are naturally selected. We have pictured camel genes as denizens of ancestral worlds, selected to survive in ancient deserts and even more ancient seas, selected to survive in companionship with compatible cartels of other camel genes. All that is true, and equivalent stories of Miocene trees and Pliocene savannahs can be told of our genes. What we must now add is that, among the worlds in which genes have survived, are virtual worlds constructed inside ancestral brains.

In the case of highly social animals like ourselves and our ancestors, our virtual worlds are, at least in part, group constructions. Especially since the invention of language and the rise of artifact and technology, our genes have had to survive in complex and changing worlds for which the most economical description we can find is shared virtual reality. It is a startling thought that, just as genes can be said to survive in deserts or forests, and just as they can be said to survive in the company of other genes in the gene pool, genes can also be said to survive in the virtual, even poetic worlds created by brains. It is to the enigma of the human brain itself that we turn in the final chapter.

12

THE BALLOON OF THE MIND

The brain is a three pound mass you can hold in your hand that can conceive of a universe a hundred billion light-years across.

MARIAN C. DIAMOND

It is a commonplace among historians of science that the biologists of any age, struggling to understand the workings of living bodies, make comparison with the advanced technology of their time. From clocks in the seventeenth century to dancing statues in the eighteenth, from Victorian heat engines to today's heat-seeking, electronically guided missiles, the engineering novelties of every age have refreshed the biological imagination. If, of all these innovations, the digital computer promises to overshadow its predecessors, the reason is simple. The computer is not just one machine. It can be swiftly reprogrammed to become any machine you like: calculator, word processor, card index, chess master, musical instrument, guess-your-weight machine, even, I regret to say, astrological soothsayer. It can simulate the weather, lemming population cycles, an ants' nest, satellite docking, or the city of Vancouver.

The brain of any animal has been described as its on-board computer. It does not work in the same way as an electronic computer. It is made from very different components. These are individually much slower, but they work in huge parallel networks so that, by some means still only partly understood, their numbers compensate for their slower speed, and brains can, in certain respects, outperform digital computers. In any case, the differences of detailed working do not disempower the metaphor. The brain is the body's on-board computer, not because of how it works but because of what it does in the life of the animal. The resemblance of role extends to many parts of the animal's economy but, perhaps most spectacularly of all, the brain simulates the world with the equivalent of virtual reality software.

It might seem a good idea, in a general way, for any animal to grow a large brain. Isn't greater computing power always likely to be an advantage? Maybe, but it has costs, too. Weight for weight, brain tissue consumes more energy than other tissues. And our big brains as babies make it quite difficult for us to be born. Our presumption that braininess must be a good thing partly grows out of vanity in our species' own hypertrophy of the brain. But it remains an interesting question why human brains have grown so especially big.

One authority has said that the evolution of the human brain over the last million years or so is 'perhaps the fastest advance recorded for any complex organ in the entire history of life'. This may be an exaggeration, but the evolution of the human brain is undeniably fast. Compared with the skulls of other apes, the modern human skull, at least the bulbous part that houses the brain, has blown up like a balloon. When we ask why this happened, it is not satisfactory to produce general reasons why having a large brain might be useful. Presumably such general benefits would apply to many kinds of animal, especially those that navigate rapidly through the complicated three-dimensional world of the forest canopy, as most primates do. A satisfying explanation will be one that tells us why one particular lineage of apes – actually, one that had left the trees – suddenly took off, leaving the rest of the primates standing.

It was once fashionable to lament – or, according to taste, gloat over – the paucity of fossils linking *Homo sapiens* to our ape ancestors. This has changed. We now have a rather good fossil series and as we go backwards in time we can trace a gradual shrinkage in braincase through various species of *Homo* to our predecessor genus *Australopithecus* whose braincase was about the same size as a modern chimpanzee's. The main difference between Lucy or Mrs Ples (famous Australopithecines) and a chimpanzee lay not in the brain at all, but in the Australopithecine habit of walking upright on two legs. Chimps only occasionally do. The blowing up of the brain balloon spanned three million years from *Australopithecus* through *Homo habilis,* then *Homo erectus,* through archaic *Homo sapiens* to modern *Homo sapiens.*

Something a bit similar seems to have happened in the growth of the computer. But, if the human brain has blown up like a balloon, the computer's progress has been more like an atom bomb. Moore's law states that the capacity of computers of a given physical size doubles every 1.5 years. (This is a modern version of the law. When Moore originally stated it more than three decades ago he was referring to transistor counts which, on his measurements, doubled every two years. Computer performance has improved even faster because transistors became faster as well as smaller and cheaper.) The late Christopher Evans, a computer-literate psychologist, put the point dramatically:

Today's car differs from those of the immediate post-war years on a number of counts. It is cheaper, allowing for the ravages of inflation, and it is more economical and efficient . . . But suppose for a moment that the automobile industry had developed at the same rate as computers and over the same period: how much cheaper and more efficient would the current models be? If you have not already heard the analogy the answer is shattering. Today you would be able to buy a Rolls-Royce for £1.35, it would do three million miles to the gallon, and it would deliver enough power to drive the Queen Elizabeth II. And if you were interested in miniaturization, you could place half a dozen of them on a pinhead. *The Mighty Micro* (1979)

Of course, things on the timescale of biological evolution inevitably happen far more slowly. One reason is that every improvement has to come about through individuals dying and rival individuals reproducing. So comparisons of absolute speed cannot be made. If we compare the brains of *Australopithecus, Homo habilis, Homo erectus* and *Homo sapiens,* we get a rough equivalent of Moore's law, slowed down by six orders of magnitude. From Lucy to *Homo sapiens,* brain size has approximately doubled every 1.5 million years. Unlike Moore's law for computers, there is no particular reason to think that the human brain will go on swelling. In order for this to happen, large-brained individuals have to have more children than small-brained individuals. It isn't obvious that this is now happening. It must have happened during our ancestral past, otherwise our brains would not have grown as they did. It also must have been true, incidentally, that braininess in our ancestors was under genetic control. If it had not been, natural selection would have had nothing to work on, and the evolutionary growth of the brain would not have occurred. For some reason, many people take grave political offence at the suggestion that some individuals are genetically cleverer than others. But this must have been the case when our brains were evolving, and there is no reason to expect that facts will suddenly change to accommodate political sensitivities.

Lots of influences have contributed to computer development which are not going to help us to understand brains. A major step was the change from the valve (vacuum tube) to the much smaller transistor, and then the spectacular and continuing miniaturization of the transistor in integrated circuits. These advances are all irrelevant to brains, because – the point deserves repetition – brains don't work electronically anyway. But there is another source of computer advancement, and this might be relevant to brains. I'll call it *self-feeding co-evolution.*

We have already met co-evolution. It means the evolving together of different organisms (as in the arms races between predators and prey), or between different parts of the same organism (the special case called co-adaptation). As another example, there are some small flies whose appearance mimics that of a jumping spider, including large dummy eyes looking straight forward like paired headlights – very unlike the compound eyes with which the flies themselves see. Real spiders are potential predators of flies of this size, but they are put off by the flies' similarity to another spider. The flies enhance the mimicry by waving their arms in ways that resemble the histrionic semaphore signals that jumping spiders use when courting their own opposite sex. In the fly, genes controlling the anatomical resemblance to spiders must have evolved together with separate genes controlling the semaphoring behaviour. This evolving together is co-adaptation.

Self-feeding is the name I am giving to any process in which 'the more you have, the more you get'. A bomb is a good example. The atomic bomb is said to depend upon a chain reaction, but the metaphor of a chain is too stately to convey what happens. When the unstable nucleus of uranium 235 breaks up, energy is released. Neutrons shooting out from the break-up of one nucleus may hit another and induce it to break up as well, but that is usually the end of the story. Most of the neutrons miss other nuclei and shoot off harmlessly into empty space, for uranium, though one of the densest of metals, is 'really', like all matter, mostly empty space. (The virtual model of metal in our brains is constructed with the persuasive illusion of dense solidity because that is the most useful internal representation of a solid for our survival purposes.) On their own scale, the atomic nuclei in a metal are far more spaced out than gnats in a swarm, and a particle expelled by one decaying atom is quite likely to have a clear run out of the swarm. If, however, you pack in a quantity (the famous 'critical mass') of uranium 235 which is just sufficient to see to it that a typical neutron expelled from any one nucleus is on average likely to hit one other nucleus before leaving the mass of metal altogether, a so-called chain reaction gets going. On average, each nucleus that splits causes another to split, there is an epidemic of atom-splitting, with an exceedingly rapid release of heat and other destructive energy, and the results are only too well known. All explosions have this same epidemic quality and, on a slower time-scale, epidemics of disease sometimes resemble explosions. They require a critical mass of susceptible victims in order to get started and, once they do get started, the more you have the more you get. This is why it is so important to vaccinate a critical proportion of the population. If fewer than the 'critical mass' remain unvaccinated, epidemics cannot take off. (This is also why it is possible for selfish free-riders to avoid being vaccinated and still benefit from the fact that most other people have been.)

In *The Blind Watchmaker* I noted a 'critical mass for explosion' principle at work in human popular culture. Many people choose to buy records, books or clothes for no better reason than that lots of other people are buying them. When a bestseller list is published, this could be seen as an objective report of purchasing behaviour. But it is more than that because the published list feeds back on people's buying behaviour and influences future sales figures. Bestseller lists are therefore, at least potentially, victims of self-feeding spirals. That's why publishers spend lots of money early in a book's career, in a strenuous attempt to nudge it over the critical threshold of the bestseller list. The hope is that then it will 'take off'. The more you have, the more you get, with the additional feature of sudden take-off, which we need for the purpose of our analogy. A dramatic example of a self-feeding spiral going in the opposite direction is the Wall Street Crash and other cases where panic selling on the stock market feeds on itself in a downward tailspin.

Evolutionary co-adaptation does not necessarily have the additional explosive property of being self-feeding. There is no reason to suppose that, in the evolution of our spider-mimicking fly, the co-adaptation of spider shape and spider behaviour was explosive. In order to be so, it is necessary that the initial resemblance, say a slight anatomical similarity to a spider, set up an *increased* pressure to mimic the spider's behaviour. This in turn fed an *even stronger* pressure to mimic the spider's shape, and so on. But, as I say, there is no reason to think it happened like this: no reason to suppose that the pressure was self-feeding and therefore increasing as it shuttled back and forth. As I explained in *The Blind Watchmaker*, it is possible that the evolution of bird of paradise tails, peacock fans and other extravagant ornaments by sexual selection is genuinely self-feeding and explosive. Here, the principle of 'the more you have, the more you get' may really apply.

In the case of the evolution of the human brain, I suspect that we are looking for something explosive, self-feeding, like the chain reaction of the atomic bomb or the evolution of a bird of paradise tail, rather than like the spider-mimicking fly. The appeal of this idea is its power to explain why, among a set of African ape species with chimpanzee-sized brains, one suddenly raced ahead of the others for no very obvious reason. It is as though a random event nudged the hominid brain over a threshold, something equivalent to a 'critical mass', and then the process took off explosively, because it was self-feeding.

What might this self-feeding process have consisted of? The conjecture I offered in my Royal Institution Christmas Lectures was 'software/hardware co-evolution'. As its name suggests, it can be explained by a computer analogy. Unfortunately for the analogy, Moore's law doesn't seem to be explained by any single self-feeding process. Integrated circuit improvement over the years seems to have been brought about by a messy collection of changes, which makes it puzzling why there is apparently steady exponential improvement. Nevertheless, there surely is some software/hardware co-evolution driving the history of computer advances. In particular, there is something corresponding to bursting through a threshold after a pent-up 'need' has been felt.

In the early days of personal computers they offered only primitive word processing software; mine didn't even 'wrap around' at the end of lines. I was then addicted to machine code programming and (I'm slightly ashamed to admit) went to the lengths of writing my own word processing software, called 'Scrivener', which I used to write *The Blind Watchmaker* – which would otherwise have been finished sooner! During the development of Scrivener, I became increasingly frustrated by the idea of using the keyboard to move the cursor around the screen. I just wanted to *point*. I toyed with using a joystick, as supplied for computer games, but couldn't work out how to do it. I overwhelmingly felt that the software I wanted to write was held up for want of a critical hardware breakthrough. Later I discovered that the device I desperately needed, but wasn't clever enough to imagine, had in fact been invented much earlier. That device was, of course, the mouse.

The mouse was a hardware advance, conceived in the 1960s by Douglas Engelbart who foresaw that it would make possible a new kind of software. This software innovation we now know, in its developed form, as the Graphical User Interface, or GUI, developed in the 1970s by the brilliantly creative team at Xerox PARC, that Athens of the modern world. It was cultivated into commercial success by Apple in 1983, then copied by other companies under names like VisiOn, GEM and – the most commercially successful today – Windows. The point of the story is that an explosion of ingenious software was, in a sense, pent up, waiting to burst on the world, but it had to wait for a crucial piece of hardware, the mouse. Subsequently, the spread of GUI software placed new demands on hardware, which had to become faster and more capacious to handle the needs of graphics. This in turn allowed a rush of more sophisticated new software, especially software capable of exploiting high-speed graphics. The software/hardware spiral continued and its latest production is the worldwide web. Who knows what may be spawned by future turns of the spiral?

Then if you look forward, it turns out the [computer] power is going to be used for a variety of things. Incremental enhancements and ease of use things, and then occasionally you go over some threshold and something new is possible. That was true with the graphical user interface. Every program got graphical and every output got graphical, that cost us vast amounts of CPU power and it was worth it . . . In fact, I have my own law of software, Nathan's Law, which is that software grows faster than Moore's Law. And that is why there is a Moore's Law.

NATHAN MYHRVOLD, Chief Technology Officer, Microsoft Corporation (1998)

Returning to the evolution of the human brain, what are we looking for to complete the analogy? A minor improvement in hardware, perhaps a slight increase in brain size, which would have gone unnoticed had it not enabled a new software technique which, in turn, unleashed a blossoming spiral of co-evolution? The new software changed the environment in which brain hardware was subject to natural selection. This gave rise to strong Darwinian pressure to improve and enlarge the hardware, to take advantage of the new software, and a self-feeding spiral was under way, with explosive results.

In the case of the human brain, what might the blossoming advance in software have been? What was the equivalent of the GUI? I'll give the clearest example I can come up with of the *kind* of thing it might have been, without for a moment committing myself to the view that this was the actual one that inaugurated the spiral. My clear example is language. Nobody knows how it began. There doesn't seem to be anything like syntax in non-human animals and it is hard to imagine evolutionary forerunners of it. Equally obscure is the origin of semantics; of words and their meanings. Sounds that mean things like 'feed me' or 'go away' are commonplace in the animal kingdom, but we humans do something quite different. Like other species, we have a limited repertoire of basic sounds, the phonemes, but we are unique in recombining those sounds, stringing them together in an indefinitely large number of combinations to mean things that are fixed only by arbitrary convention. Human language is open-ended in its semantics: phonemes can be recombined to concoct an indefinitely expanding dictionary of words. And it is open-ended in its syntax, too: words can be recombined in an indefinitely large number of sentences by recursive embedment: 'The man is coming. The man who caught the leopard is coming. The man who caught the leopard which killed the goats is coming. The man who caught the leopard which killed the goats who give us our milk is coming.' Notice how the sentence grows in the middle while the ends – its fundamentals – stay the same. Each of the embedded subordinate clauses is capable of growing in the same way, and there is no limit to the permissible growth. This kind of potentially infinite enlargement, which is suddenly made possible by a single syntactic innovation, seems to be unique to human language.

Nobody knows whether our ancestors' language went through a prototype stage with a small vocabulary and a simple grammar before gradually evolving to the present point where all the thousands of languages in the world are very complex (some say they are all exactly *equally* complex, but that sounds too ideologically perfect to be wholly plausible). I am biased towards thinking that it was gradual, but it is not quite obvious that it had to be. Some people think it began suddenly, more or less literally invented by a single genius in a particular place at a particular time. Whether it was gradual or sudden, a similar story of software/hardware co-evolution could be told. A social world in which there is language is a completely different kind of social world from one in which there is not. The selection pressures on genes will never be the same again. The genes find themselves in a world that is more dramatically different than if an ice age had suddenly struck or some terrible new predator had suddenly arrived in the land. In the new social world where language first burst on the scene, there must have been dramatic natural selection in favour of individuals genetically equipped to exploit the new ways. It is reminiscent of the conclusion of the previous chapter, in which I spoke of genes being selected to survive in the virtual worlds constructed socially by brains. It is almost impossible to overestimate the advantages that could have been enjoyed by individuals able to excel in taking advantage of the new world of language. It is not just that brains became bigger to cope with managing language itself. It is also that the whole world in which our ancestors lived was transformed as a consequence of the invention of speaking.

But I used the example of language just to make the idea of software/hardware co-evolution plausible. It may not have been language that pushed the human brain over its critical threshold for inflation, although I have a hunch that it played an important role. It is controversial whether the sound-modulating hardware in the throat was capable of language at the time when the brain began to swell up. There is some fossil evidence to suggest that our likely ancestors *Homo habilis* and *Homo erectus,* because of their relatively undescended larynx, probably were not capable of articulating the full range of vowel sounds that modern throats put at our disposal. Some people take this as indicating that language itself arrived late in our evolution. I think this a rather unimaginative conclusion. If there was software/hardware co-evolution, the brain is not the only hardware that we should expect to have improved in the spiral. The vocal apparatus, too, would have evolved in parallel, and the evolutionary descent of the larynx is one of the hardware changes that language itself would drive. Poor vowels are not the same thing as no vowels at all. Even if *Homo erectus* speech sounded monotonous by our exacting standards, it could still have served as the arena for the evolution of syntax, semantics and the self-feeding descent of the larynx itself. *Homo erectus*, incidentally, conceivably made boats as well as fire; we should not underestimate them.

Setting language on one side for a moment, what other software innovations might have nudged our ancestors over the critical threshold and initiated the co-evolutionary escalation? Let me suggest two that could have arisen naturally from our ancestors' evolving fondness for meat and hunting. Agriculture is a recent invention. Most of our hominid ancestors have been hunter gatherers. Those who still subsist from this ancient way of life are often formidable trackers. They can read patterns of footprints, disturbed vegetation, dung deposits and traces of hair to build up a detailed picture of events over a wide area. A pattern of footprints is a graph, a map, a symbolic representation of a series of incidents in animal behaviour. Remember our hypothetical zoologist, whose ability to reconstruct past environments by reading an animal's body and its DNA justified the statement that an animal is a model of its environment? Mightn't we say something similar of an expert !Kung San tracker, who has only to read footprints in the Kalahari dirt to reconstruct a detailed pattern, description, or model of animal behaviour in the recent past? Properly read, such spoors amount to maps and pictures, and it seems to me plausible that the ability to read such maps and pictures might have arisen in our ancestors before the origin of speech in words.

Suppose that a band of *Homo habilis* hunters needed to plan a cooperative hunt. In a remarkable and chilling 1992 television film, *Too Close for Comfort*, David Attenborough shows modern chimpanzees executing what seems to be a carefully planned and successful drive and ambush of a colobus monkey, which they then tear to pieces and eat. There is no reason to think that the chimpanzees communicated any detailed plan to each other before beginning the hunt, but every reason to think that *habilis* might have benefited from some such communication if it could have been achieved. How might such communication have developed?

Suppose that one of the hunters, whom we can think of as a leader, has a plan to ambush an eland and he wishes to convey the plan to his colleagues. No doubt he could mime the behaviour of the eland, perhaps donning an eland skin for the purpose, as hunting peoples do today for ritual or entertainment purposes. And he could mime the actions he wants his hunters to perform: studied exaggeration of stealth in the stalk; noisy conspicuousness in the drive; sudden startle in the final ambush. But there is more that he could do, and in this he would resemble any modern army officer. He could point out objectives and planning manoeuvres on a *map* of the area.

Our hunters, we may suppose, are all expert trackers, with a feel for the layout, in two-dimensional space, of footsteps and other traces: a spatial expertise which may have been beyond anything we (unless we happen to be !Kung San hunters ourselves) can easily imagine. They are all fully accustomed to the idea of following a trail, and imagining it laid out on the ground as a life-size map and a temporal graph of the movements of an animal. What could be more natural than for the leader to seize a stick and draw in the dust a scale model of just such a temporal picture: a map of movement over a surface? The leader and his hunters are fully used to the idea that a series of hoofprints indicate the flow of wildebeests along the muddy bank of a river. Why should he not draw a line indicating the flow of the river itself on a scale map in the dust? Accustomed as they all are to following human footprints from their own home cave to the river, why would the leader not point on his map to the position of the cave in relation to the river? Moving around the map with his stick, the hunter could indicate the direction of approach by the eland, the angle of his proposed drive, the location of the ambush: indicate them literally by drawing in the sand.

Could something like this have been how the notion of a scaled-down representation in two dimensions was born – as a natural generalization of the important skill of reading animal footprints? Maybe the idea of drawing the likeness of animals themselves arose from the same source. The imprint in mud of a wildebeest hoof is obviously a negative image of the real thing. The fresh paw mark of a lion must have aroused fear. Did it also engender in a blinding flash the realization that one could draw a representation of a part of an animal – and hence, by extrapolation, of the whole animal? Perhaps the blinding flash that led to the first drawing of a whole animal came from the imprint of a whole corpse, dragged out of mud which had baked hard around it. Or a less distinct image in the grass could easily have been fleshed out by the mind's own virtual reality software.

> *Because the mountain grass*
> *Cannot but keep the form*
> *Where the mountain hare has lain.*
>
> W. B. YEATS, 'Memory' (1919)

Representational art of all kinds (and probably non-representational art, too) depends upon noticing that something can be made to stand for something else and that this may assist thought or communication. The analogies and metaphors that underlie what I have been calling poetic science – good and bad – are other manifestations of the same human faculty of symbol-making. Let's recognize a continuum, which could represent an evolutionary series. At one end of the continuum we allow things to stand for other things that they resemble – as in cave paintings of buffaloes. At the other end are symbols which do not obviously resemble the things that they stand for – as in the word 'buffalo', which means what it does only because of an arbitrary convention which all English speakers respect. The intermediate stages along the continuum may, as I said, represent an evolutionary progression. We may never know how it began. But perhaps my story of the footprints represents the *kind* of insight that might have been involved when people first began to think by analogy, and hence realize the possibility of semantic representation. Whether or not it gave birth to semantics, my tracker map joins language as my second suggestion for a software innovation that may have triggered the co-evolutionary spiral that drove the expansion of our brain. Could it have been the drawing of maps that boosted our ancestors beyond the critical threshold which the other apes just failed to cross?

My third possible software innovation is inspired by a suggestion made by William Calvin. He proposed that ballistic movements, such as throwing projectiles at a distant target, make special computational demands on nervous tissue. His idea was that the conquering of this particular problem, perhaps originally for purposes of hunting, equipped the brain to do lots of other important things as a by-product.

On a shingle beach, Calvin was amusing himself by tossing stones at a log and the action inadvertently launched (the metaphor is no accident) a productive train of thought. What kind of computation must the brain be doing when we throw something at a target, as our ancestors must increasingly have done while they evolved the hunting habit? One crucial component of an accurate throw is timing. Whichever arm action you favour, whether underarm lobbing, overarm bowling or throwing, or wristy flicking, the exact moment at which you release your projectile makes all the difference. Think about the overarm action of a bowler in cricket (bowling differs from baseball pitching in that the arm must remain straight, and this makes it easier to think about). If you release the ball too soon, it flies over the batsman's head. If you let go too late, it digs into the ground. How does the nervous system achieve the feat of releasing the projectile at exactly the right moment, tailored to the speed of arm movement? Unlike a lunge with a sword, in which you might steer your aim all the way to the target, bowling or throwing is ballistic. The projectile leaves your hand and is then beyond your control. There are other skilled movements, like hammering a nail, which are effectively ballistic, even if the tool or weapon doesn't leave your hand. All the computation has to be done in advance: 'dead reckoning'.

One way to solve the release timing problem when throwing a stone or a spear would be to compute the necessary contractions of individual muscles on the fly, while the arm was in motion. Modern digital computers would be capable of this feat, but brains are too slow. Calvin reasoned instead that nervous systems, being slow, would be better off with a buffer store of rote commands to the muscles. The whole sequence of bowling a cricket ball, or throwing a spear, is programmed in the brain as a pre-recorded list of individual muscle twitch commands, packed away in the order they are to be released.

Obviously, more distant targets are harder to hit. Calvin dusted off his physics textbooks and worked out how to calculate the decreasing 'launch window' as you try to maintain accuracy for longer and longer throws. Launch window is space jargon. Rocket scientists (that proverbially gifted profession) calculate the window of opportunity during which they must launch a spacecraft if they are to hit, say, the moon. Fire too soon, or too late, and you miss. Calvin worked out that for a rabbit-sized target four metres away, his launch window was about 11 milliseconds wide. If he released his stone too soon, it overshot the rabbit. If he held on too long, his stone fell short. The difference between two short and too long was a mere 11 milliseconds, about a hundredth of a second. Being an expert in the timings of nerve cells, this bothered Calvin, because he knew that the normal margin of error of a nerve cell is greater than the launch window. Yet he also knew that good human throwers are capable of hitting such a target at this distance, even while running. I myself have never forgotten the spectacle of my Oxford contemporary the Nawab of Pataudi (one of India's greatest cricketers, even after losing one eye) fielding for the university and throwing the ball with devastating speed and accuracy at the wicket, again and again, even while running at a speed that visibly intimidated the batsmen while raising the game of his team.

Calvin had a mystery to solve. How do we throw so well? The answer, he decided, must lie in the law of large numbers. No one timing circuit can achieve the accuracy of a !Kung hunter throwing a spear, or a cricketer throwing a ball. There must be lots of timing circuits working in parallel, their effects being averaged to reach the final decision of when to release the projectile. And now comes the point. Having developed a population of timing and sequencing circuits for one purpose, why not turn them to other ends? Language itself relies upon precise sequencing. So does music, dancing, even thinking out plans for the future. Could throwing have been the forerunner of foresight itself? When we throw our mind forward in imagination, are we doing something almost literal as well as metaphorical? When the first word was uttered, somewhere in Africa, did the speaker imagine himself throwing a missile from his mouth to his intended hearer?

My fourth candidate for software that partakes in software/hardware co-evolution is the 'meme', the unit of cultural inheritance. We've already hinted at it when discussing the epidemic-style 'take-off' of bestsellers. I here draw upon books of my colleagues Daniel Dennett and Susan Blackmore, who have been among several constructive memetic theorists since the word was first coined in 1976. Genes are replicated, copied from parent to offspring down the generations. A meme is, by analogy, anything that replicates itself from brain to brain, via any available means of copying. It is a matter of dispute whether the resemblance between gene and meme is good scientific poetry or bad. On balance, I still think it is good, although if you look the word up on the worldwide web you'll find plenty of examples of enthusiasts getting carried away and going too far. There even seems to be some kind of religion of the meme starting up – I find it hard to decide whether it is a joke or not.

My wife and I both occasionally suffer from sleeplessness when our minds are taken over by a tune which repeats itself over and over in the head, relentlessly and without mercy, all through the night. Certain tunes are especially bad culprits, for example Tom Lehrer's 'Masochism Tango'. This is not a melody of any great merit (unlike the words, which are brilliantly rhymed), but it is almost impossible to shake off once it gains a hold. We now have a pact that, if we have one of the danger tunes on the brain during the day (Lennon and McCartney are other prime culprits), we shall under no circumstances sing or whistle it near bedtime, for fear of infecting the other. This notion that a tune in one brain can 'infect' another brain is pure meme talk.

The same thing can happen when one is awake. Dennett tells the following anecdote in *Darwin's Dangerous Idea* (1995):

The other day, I was embarrassed – dismayed – to catch myself walking along humming a melody to myself. It was not a theme of Haydn or Brahms or Charlie Parker or even Bob Dylan: I was energetically humming 'It takes two to tango' – a perfectly dismal and entirely unredeemed bit of chewing gum for the ears that was unaccountably popular sometime in the 1950s. I am sure I have never in my life chosen this melody, esteemed this melody, or in any way judged it to be better than silence, but there it was, a horrible musical virus, at least as robust in the meme pool as any melody I actually esteem. And now, to make matters worse, I have resurrected the virus in many of you, who will no doubt curse me in days to come when you find yourself humming, for the first time in over thirty years, that boring tune.

For me, the maddening jingle is just as often not a tune but an endlessly repeated phrase, not a phrase with any obvious significance, just a fragment of language that I, or somebody else, has perhaps said at some point during the day. It isn't clear why a particular phrase or tune is chosen but, once there, it is extremely hard to shift. It goes on endlessly rehearsing itself. In 1876 Mark Twain wrote a short story, 'A Literary Nightmare', about his mind being taken over by a ridiculous fragment of versified instruction to a bus conductor with a ticket machine, of which the refrain was 'Punch in the presence of the passenjare'.

Punch in the presence of the passenjare
Punch in the presence of the passenjare

It has a mantra-like rhythm and I almost dared not quote it for fear of infecting you. I had it going round in my own head for a whole day after reading Mark Twain's story. Twain's narrator finally liberated himself by passing it on to the vicar, who in turn was driven demented. This 'Gadarene swine' aspect of the story – the idea that when you pass a meme to somebody else you thereby lose it – is the only part that does not ring true. Just because you infect somebody else with a meme, does not mean you cleanse your brain of it.

Memes can be good ideas, good tunes, good poems, as well as drivelling mantras. Anything that spreads by imitation, as genes spread by bodily reproduction or by viral infection, is a meme. The chief interest of them is that there is at least the theoretical possibility of a true Darwinian selection of memes, to parallel the familiar selection of genes. Those memes that spread do so because they are good at spreading. Dennett's relentless jingle, like mine and my wife's, was a tango. Is there something insidious about the tango rhythm? Well, we need further evidence. But the general idea that some memes may be more infective than others because of their inherent properties is reasonable enough.

As with genes, we can expect the world to become filled with memes that are good at the art of getting themselves copied from brain to brain. We can notice that some memes, like Mark Twain's jingle, have this property as a matter of fact, though without being able to analyse what gives it to them. It is enough that memes vary in their infectivity for Darwinian selection to get going. Sometimes we can work out what it is that a meme has that helps it to spread. Dennett notes that the conspiracy theory meme has a built-in response to the objection that there is no good evidence for the conspiracy: 'Of course not – that's how powerful the conspiracy is!'

Genes will spread by reason of pure parasitic effectiveness, as in a virus. We may think this spreading for the sake of spreading rather futile, but nature is not interested in our judgements, of futility or of anything else. If a piece of code has what it takes, it spreads and that's that. Genes can also spread for what we think of as a more 'legitimate' reason, say, because they improve the acuity of a hawk's eyesight. They are the ones that first occur to us when we think of Darwinism. In *Climbing Mount Improbable* I explained that an elephant's DNA and a virus's are both 'Copy Me' programmes. The difference is that that one of them has an almost fantastically large digression: 'Copy me by building an elephant first.' But both kinds of program spread because, in their different ways, they are good at spreading. The same is true for memes. Jingling tangos survive in brains, and infect other brains, for reasons of pure parasitic effectiveness. They are near the virus end of the spectrum. Great ideas in philosophy, brilliant insights in mathematics, clever techniques for tying knots or fashioning pots, survive in the meme pool for reasons that are closer to the 'legitimate' or 'elephant' end of our Darwinian spectrum.

Memes could not spread but for the biologically valuable tendency of individuals to imitate. There are plenty of good reasons why imitation should have been favoured by conventional natural selection working on genes. Individuals that are genetically predisposed to imitate enjoy a fast track to skills that may have taken others a long time to build up. One of the finest examples is the spread of the habit of opening milk bottles among tits (European equivalent of American chickadees). Milk is delivered in bottles very early to British doorsteps and it usually sits there for a while before being taken in. A small bird is capable of pecking through the lid, but it is not an obvious thing for a bird to do. What happened was that a series of epidemics of bottletop raiding among blue tits spread outwards from discrete geographical foci in Britain. Epidemic is exactly the right word. The zoologists James Fisher and Robert Hinde were able to document the spread of the habit in the 1940s as it radiated outwards by imitation from the focal points where it started, presumably discovered by a few isolated birds: islands of inventiveness and founders of meme epidemics.

Similar stories can be told of chimpanzees. Fishing for termites by poking twigs into a mound is learned by imitation. So is the skill of cracking nuts with stones on a log or stone anvil, which occurs in certain local areas of west Africa but not others. Our hominid ancestors surely learned vital skills by imitating each other. Among surviving tribal groups, stone toolmaking, weaving, techniques for fishing, thatching, pottery, firemaking, cooking, smithwork, all these skills are learned by imitation. Lineages of masters and apprentices are the memetic equivalent of genetic ancestor/descendant lines. The zoologist Jonathan Kingdon has suggested that some of our ancestors' skills began when humans imitated other species. For example, spider webs may have inspired the invention of fishing nets and of string or twine, weaver bird nests the invention of knots or thatching.

Memes, unlike genes, don't seem to have clubbed together to build large 'vehicles' – bodies – for their joint housing and survival. Memes rely on the vehicles built by genes (unless, as has been suggested, you count the Internet as a meme vehicle). But memes manipulate the behaviour of living bodies no less effectively for that. The analogy between genetic and memetic evolution starts to get interesting when we apply our lesson of 'the selfish cooperator'. Memes, like genes, survive in the presence of certain other memes. A mind can become prepared, by the presence of certain memes, to be receptive to particular other memes. Just as a species gene pool becomes a cooperative cartel of genes, so a group of minds – a 'culture', a 'tradition' – becomes a cooperative cartel of memes, a memeplex, as it has been called. As in the case of genes, it is a mistake to see the whole cartel as a unit being selected as a single entity. The right way to see it is in terms of mutually assisting memes, each providing an environment which favours the others. Whatever may be the limitations of the meme theory, I think this one point, that a culture or a tradition, a religion or a political complexion grows up according to the model of 'the selfish cooperator' is probably at least an important part of the truth.

Dennett vividly evokes the image of the mind as a seething hotbed of memes. He even goes so far as to defend the hypothesis that 'Human consciousness is *itself* a huge complex of memes . . .' He does this, along with much else, persuasively and at length, in his book *Consciousness Explained* (1991). I cannot possibly summarize the intricate series of arguments in that book, and will content myself with one more characteristic quotation:

The haven all memes depend on reaching is the human mind, but a human mind itself is an artifact created when memes restructure a human brain in order to make it a better habitat for memes. The avenues for entry and departure are modified to suit local conditions, and strengthened by various artificial devices that enhance fidelity and prolixity of replication: native Chinese minds differ dramatically from native French minds, and literate minds differ from illiterate minds. What memes provide in return to the organisms in which they reside is an incalculable store of advantages – with some Trojan horses thrown in for good measure . . . But if it is true that human minds are themselves to a very great degree the creations of memes, then we cannot sustain the polarity of vision we considered earlier; it cannot be 'memes versus us,' because earlier infestations of memes have already played a major role in determining who or what we are.

There is an ecology of memes, a tropical rainforest of memes, a termite mound of memes. Memes don't only leap from mind to mind by imitation, in culture. That is just the easily visible tip of the iceberg. They also thrive, multiply and compete within our minds. When we announce to the world a good idea, who knows what subconscious quasi-Darwinian selection has gone on behind the scenes inside our heads? Our minds are invaded by memes as ancient bacteria invaded our ancestors' cells and became mitochondria. Cheshire Cat-like, memes merge into our minds, even become our minds, just as eucaryotic cells are colonies of mitochondria, chloroplasts and other bacteria. This sounds like a perfect recipe for co-evolutionary spirals and the enlargement of the human brain, but specifically what drives the spiral? Where lies the self-feeding, the element of 'the more you have, the more you get'?

Susan Blackmore tackles this question, by asking another: 'Whom should you imitate?' The individuals who are best at the skill in question, certainly, but there is a more general answer to the question. Blackmore suggests that you should choose to imitate the best imitators – they are likely to have picked up the best skills. And her next question, 'With whom do you mate?' is answered in a similar way. You mate with the best imitators of the trendiest memes. So, not only are memes selected for the ability to spread themselves, genes are selected in ordinary Darwinian selection for their ability to make individuals that are good at spreading memes. I do not wish to steal Doctor Blackmore's thunder, for I have been privileged to see an advance draft of her book, *The Meme Machine* (1999). I will simply note that here we have software/hardware co-evolution. The genes build the hardware. The memes are the software. The co-evolution is what may have driven the inflation of the human brain.

I said that I'd return to the illusion of the 'little man in the brain'. Not to solve the problem of consciousness, which is way beyond my capacity, but to make another comparison between memes and genes. In *The Extended Phenotype*, I argued against taking the individual organism for granted. I didn't mean individual in the conscious sense but in the sense of a single, coherent body surrounded by a skin and dedicated to a more or less unitary purpose of surviving and reproducing. The individual organism, I argued, is not fundamental to life, but something that emerges when genes, which at the beginning of evolution were separate, warring entities, gang together in cooperative groups, as 'selfish cooperators'. The individual organism is not exactly an illusion. It is too concrete for that. But it is a secondary, derived phenomenon, cobbled together as a consequence of the actions of fundamentally separate, even warring, agents. I shan't develop the idea but just float, following Dennett and Blackmore, the idea of a comparison with memes. Perhaps the subjective 'I', the person that I feel myself to be, is the same kind of semi-illusion. The mind is a collection of fundamentally independent, even warring, agents. Marvin Minsky, the father of artificial intelligence, called his 1985 book *The Society of Mind*. Whether or not these agents are to be identified with memes, the point I am now making is that the subjective feeling of 'somebody in there' may be a cobbled, emergent, semi-illusion analogous to the individual body emerging in evolution from the uneasy cooperation of genes.

But that was an aside. I have been looking for software innovations that might have launched a self-feeding spiral of hardware/software co-evolution to account for the inflation of the human brain. I have so far mentioned language, map reading, throwing and memes. Another possibility is sexual selection, which I introduced as an analogy to explain the principle of explosive co-evolution, but could it actually have driven the inflation of the human brain? Did our ancestors impress their mates by a sort of mental peacock's tail? Was larger brain hardware favoured because of its ostentatious software manifestations, perhaps as the ability to remember the steps of a formidably complicated ritual dance? Perhaps.

Many people will find language itself the most persuasive, as well as the clearest candidate for a software trigger of brain expansion, and I'd like to come back to it from another point of view. Terrence Deacon, in *The Symbolic Species* (1997), has a meme-like approach to language:

It is not too far-fetched to think of languages a bit as we think of viruses, neglecting the difference in constructive versus destructive effects. Languages are inanimate artifacts, patterns of sounds and scribblings on clay or paper, that happen to get insinuated into the activities of human brains which replicate their parts, assemble them into systems, and pass them on. The fact that the replicated information that constitutes a language is not organized into an animate being in no way excludes it from being an integrated adaptive entity evolving with respect to human hosts.

Deacon goes on to prefer a 'symbiotic' rather than a virulently parasitic model, drawing the comparison again with mitochondria and other symbiotic bacteria in cells. Languages evolve to become good at infecting child brains. But the brains of children, those mental caterpillars, also evolve to become good at being infected by language: co-evolution yet again.

C. S. Lewis, in 'Bluspels and Flalansferes' (1939), reminds us of the philologist's aphorism that our language is full of dead metaphors. In his 1844 essay 'The Poet', the philosopher and poet Ralph Waldo Emerson said, 'Language is fossil poetry.' If not all of our words, certainly a great number of them, began as metaphors. Lewis mentions 'attend' as having once meant 'stretch'. If I attend to you, I stretch my ears towards you. I 'grasp' your meaning as you 'cover' your topic and 'drive home' your 'point'. We 'go into' a subject, 'open up' a 'line' of thought. I have deliberately chosen cases whose metaphoric ancestry is recent and therefore accessible. Philological scholars will delve deeper (see what I mean?) and show that even words whose origins are less obvious were once metaphors, perhaps in a dead (get it?) language. The word language itself comes from the Latin for tongue.

I have just bought a dictionary of contemporary slang because I was disconcerted to be told by American readers of the typescript of this book that some of my favourite English words would not be understood across the Atlantic. 'Mug', for instance, meaning fool, dupe or patsy, is not understood there. In general I have been reassured to find from the dictionary how many slang words are actually universal in the English-speaking world. But I have been more intrigued at the astonishing creativeness of our species in inventing an endless supply of new words and usages. 'Parallel parking' or 'getting your plumbing snaked' for copulation; 'idiot box' for television; 'park a custard' for vomit; 'Christmas on a stick' for a conceited person; 'nixon' for a fraudulent deal; 'jam sandwich' for a police car; these slang expressions represent the cutting edge of an astonishing richness of semantic innovation. And they perfectly illustrate C. S. Lewis's point. Is this how all our words got their start?

As with the 'footprint maps', I wonder whether the ability to see analogies, the ability to express meanings in terms of symbolic resemblances to other things, may have been the crucial software advance that propelled human brain evolution over the threshold into a co-evolutionary spiral. In English we use the word 'mammoth' as an adjective, synonymous with very large. Could our ancestors' breakthrough into semantics have come when some pre-sapient poetic genius, struggling to convey the idea of 'large' *in some quite different context* hit upon the idea of imitating, or drawing, a mammoth? Could that have been the kind of software advance that nudged humanity into an explosion of software/hardware co-evolution? Perhaps not this particular example, because large size is too easily conveyed by the universal hand gesture beloved of boastful anglers. But even that is a software advance over chimpanzee communication in the wild. Or how about imitating a gazelle to *mean* the delicate, shy grace of a girl, in a Pliocene anticipation of Yeats's 'Two girls, both beautiful, one a gazelle'? How about sprinkling water from a gourd to mean not just rain, which is almost too obvious, but tears when trying to convey sadness? Could our remote *habilis* or *erectus* ancestors have imagined – and momentously discovered the means to express – an image like the 'sobbing rain' of John Keats? (Though, to be sure, tears themselves are an unsolved evolutionary mystery.)

However it began, and whatever its role in the evolution of language, we humans, uniquely among animalkind, have the poet's gift of metaphor: of noticing when things are like other things and using the relation as a fulcrum for our thoughts and feelings. This is an aspect of the gift of imagining. Perhaps *this* was the key software innovation that triggered our co-evolutionary spiral. We could think of it as a key advance in the world-simulating software that was the subject of the previous chapter. Perhaps it was the step from constrained virtual reality, where the brain simulates a model of what the sense organs are telling it, to unconstrained virtual reality, in which the brain simulates things that are not actually there at the time – imagination, daydreaming, 'what if?' calculations about hypothetical futures. And this, finally, brings us back to poetic science and the dominant theme of the whole book.

We can take the virtual reality software in our heads and emancipate it from the tyranny of simulating only utilitarian reality. We can imagine worlds that might be, as well as those that are. We can simulate possible futures as well as ancestral pasts. With the aid of external memories and symbol-manipulating artifacts – paper and pens, abacuses and computers – we are in a position to construct a working model of the universe and run it in our heads before we die.

We can get outside the universe. I mean in the sense of putting a model of the universe *inside* our skulls. Not a superstitious, small-minded, parochial model filled with spirits and hobgoblins, astrology and magic, glittering with fake crocks of gold where the rainbow ends. A big model, worthy of the reality that regulates, updates and tempers it; a model of stars and great distances, where Einstein's noble spacetime curve upstages the curve of Yahweh's covenantal bow and cuts it down to size; a powerful model, incorporating the past, steering us through the present, capable of running far ahead to offer detailed constructions of alternative futures and allow us to choose.

Only human beings guide their behaviour by a knowledge of what happened before they were born and a preconception of what may happen after they are dead; thus only humans find their way by a light that illuminates more than the patch of ground they stand on.

P. B. and J. S. MEDAWAR, *The Life Science* (1977)

The spotlight passes but, exhilaratingly, before doing so it gives us time to comprehend something of this place in which we fleetingly find ourselves and the reason that we do so. We are alone among animals in foreseeing our end. We are also alone among animals in being able to say before we die: Yes, this is why it was worth coming to life in the first place.

> *Now more than ever seems it rich to die,*
> *To cease upon the midnight with no pain,*
> *While thou art pouring forth thy soul abroad*
> *In such an ecstasy!*
>
> JOHN KEATS, 'Ode to a Nightingale' (1820)

A Keats and a Newton, listening to each other, might hear the galaxies sing.

SELECTED BIBLIOGRAPHY

1. Alvarez, W. (1997) *T. rex and the Crater of Doom.* Princeton, NJ: Princeton University Press.
2. Appleyard, B. (1992) *Understanding the Present.* London: Picador.
3. Asimov, I. (1979) *The Book of Facts, Volume 2.* London: Hodder & Stoughton.
4. Atkins, P. W. (1984) *The Second Law.* New York: Scientific American.
5. Atkins, P. W. (1992) *Creation Revisited.* Oxford: W. H. Freeman.
6. Attneave, F. (1954) Informational aspects of visual perception. *Psychological Reviews*, **61**, 183–93.
7. Barkow, J. H., Cosmides, L., & Tooby, J. (1992) *The Adapted Mind.* New York: Oxford University Press.
8. Barlow, H. B. (1963) The coding of sensory messages. In W. H. Thorpe & O. L. Zangwill (eds.), *Current Problems in Animal Behaviour.* Cambridge: Cambridge University Press, 331–60.
9. Barrow, J. D. (1998) *Impossibility: The Limits of Science and the Science of Limits.* Oxford: Oxford University Press.
10. Blackmore, S. (1999) *The Meme Machine.* Oxford: Oxford University Press.
11. Bodmer, W., & McKie, R. (1994) *The Book of Man: The Quest to Discover Our Genetic Heritage.* London: Little, Brown.
12. Bragg, M. (1998) *On Giants' Shoulders.* London: Hodder & Stoughton.
13. Brockman, J. (1995) *The Third Culture.* New York: Simon & Schuster.
14. Brockman, J., & Matson, K. (eds.) (1996) *How Things Are: A Science Toolkit for the Mind.* London: Phoenix.

15. Cairns-Smith, A. G. (1996) *Evolving the Mind.* Cambridge: Cambridge University Press.

16. Calvin, W. H. (1989) *The Cerebral Symphony.* New York: Bantam Books.

17. Calvin, W. H. (1996) *How Brains Think.* London: Weidenfeld & Nicolson.

18. Carey, J. (1995) *The Faber Book of Science.* London: Faber & Faber.

19. Cartmill, M. (1998) Oppressed by evolution. *Discover,* March, 78–83.

20. Clarke, A. C. (1982) *Profiles of the Future.* London: Victor Gollancz.

21. Conway Morris, S. (1998) *The Crucible of Creation.* Oxford: Oxford University Press.

22. Cook, E. (1990) *John Keats.* Oxford: Oxford University Press.

23. Craik, K. J. W. (1943) *The Nature of Explanation.* London: Cambridge University Press.

24. Crick, F. (1994) *The Astonishing Hypothesis.* New York: Scribners.

25. Cronin, H. (1991) *The Ant and the Peacock.* Cambridge: Cambridge University Press.

26. Darwin, C. (1859) *On the Origin of Species.* London (1968): Penguin Books.

27. Davies, N. B. (1992) *Dunnock Behaviour and Social Evolution.* Oxford: Oxford University Press.

28. Dawkins, M. S. (1993) *Through Our Eyes Only?* Oxford: W. H. Freeman.

29. Dawkins, R. (1982) *The Extended Phenotype.* Oxford: Oxford University Press.

30. Dawkins, R. (1986) *The Blind Watchmaker.* London: Penguin Books.

31. Dawkins, R. (1989) *The Selfish Gene.* Second Edition. Oxford: Oxford University Press.

32. Dawkins, R. (1995) *River Out of Eden*. London: Weidenfeld & Nicolson.

33. Dawkins, R. (1996) *Climbing Mount Improbable*. New York: Norton.

34. Dawkins, R. (1998) The values of science and the science of values. In J. Ree & C. W. C. Williams (eds.), *The Values of Science: The Oxford Amnesty Lectures 1997*. Boulder, Colo.: Westview Press.

35. de Waal, F. (1996) *Good Natured*. Cambridge, Mass.: Harvard University Press.

36. Deacon, T. (1997) *The Symbolic Species*. London: Allen Lane.

37. Dean, G., Mather, A., & Kelly, I. W. (1996) Astrology. In G. Stein (ed.), *The Encyclopedia of the Paranormal*. Amherst, NY: Prometheus Books, 47–99.

38. Dennett, D. C. (1991) *Consciousness Explained*. Boston: Little, Brown.

39. Dennett, D. C. (1995) *Darwin's Dangerous Idea*. New York: Simon & Schuster.

40. Deutsch, D. (1997) *The Fabric of Reality*. London: Allen Lane.

41. Dunbar, R. (1995) *The Trouble with Science*. London: Faber & Faber.

42. Durham, W. H. (1991) *Coevolution: Genes, Culture and Human Diversity*. Stanford: Stanford University Press.

43. Dyson, F. (1997) *Imagined Worlds*. Cambridge, Mass.: Harvard University Press.

44. Eddington, A. (1928) *The Nature of the Physical World*. Cambridge: Cambridge University Press.

45. Ehrenreich, B., & McIntosh, J. (1997) The new creationism. *The Nation*, 9 June.

46. Einstein, A. (1961) *Relativity: The Special and the General Theory*. New York: Bonanza Books.

47. Eiseley, L. (1982) *The Firmament of Time*. London: Victor Gollancz.

48. Evans, C. (1979) *The Mighty Micro*. London: Victor Gollancz.

49. Feller, W. (1957) *An Introduction to Probability Theory and its Applications*. New York: Wiley International Edition.

50. Feynman, R. P. (1965) *The Character of Physical Law*. London: Penguin.

51. Feynman, R. P. (1998) *The Meaning of It All*. London: Penguin Books.

52. Fisher, J., & Hinde, R. A. (1949) The opening of milk bottles by birds. *British Birds*, **42**, 347–57.

53. Ford, E. B. (1975) *Ecological Genetics*. London: Chapman & Hall.

54. Frazer, J. G. (1922) *The Golden Bough*. London: Macmillan.

55. Freeman, D. (1998) *The Fateful Hoaxing of Margaret Mead: An Historical Analysis of Her Samoan Researches*. Boulder, Colo.: Westview Press.

56. Fruman, N. (1971) *Coleridge, the Damaged Archangel*. London: Allen & Unwin.

57. Good, I. J. (1995) When batterer turns murderer. *Nature*, **375**, 541.

58. Gould, S. J. (1977) Eternal metaphors of paleontology. In A. Hallam (ed.), *Patterns of Evolution, As Illustrated by the Fossil Record*. Amsterdam: Elsevier, 1–26.

59. Gould, S. J. (1989) *Wonderful Life: The Burgess Shale and the Nature of History*. London: Hutchinson Radius.

60. Gregory, R. L. (1981) *Mind in Science: A History of Explanations in Psychology and Physics*. London: Weidenfeld & Nicolson.

61. Gregory, R. L. (1998) *Eye and Brain*. Fifth Edition. Oxford: Oxford University Press.

62. Gribbin, J., & Cherfas, J. (1982) *The Monkey Puzzle*. London: The Bodley Head.

63. Gross, P. R., & Levitt, N. (1994) *Higher Superstition: The Academic Left and its Quarrels with Science*. Baltimore: Johns Hopkins University Press.

64. Hamilton, W. D. (1996) *Narrow Roads of Gene Land: The*

Collected Papers of W. D. Hamilton. Vol. 1. Evolution of Social Behaviour. Oxford: W. H. Freeman/Spektrum.

65. Hardin, C. L. (1988) *Color for Philosophers: Unweaving the Rainbow.* Indianapolis: Hackett.
66. Heath-Stubbs, J., & Salman, P. (eds.) (1984) *Poems of Science.* London: Penguin Books.
67. Hoffmann, B. (1973) *Einstein.* London: Paladin.
68. Hölldobler, B., & Wilson, E. O. (1990) *The Ants.* Berlin: Springer-Verlag.
69. Hoyle, F. (1966) *Man in the Universe.* New York: Columbia University Press.
70. Hume, D. (1748) *An Enquiry Concerning Human Understanding.* 'Of Miracles'. Oxford: Oxford University Press (ed. L. A. Selby-Bigge, 1902).
71. Humphrey, N. (1995) *Soul Searching.* London: Chatto & Windus.
72. Humphrey, N. (1998) What shall we tell the children? In J. Ree & C. W. C. Williams (eds.), *The Values of Science: The Oxford Amnesty Lectures 1997.* Boulder, Colo.: Westview Press.
73. Huxley, T. H. (1894) *Collected Essays.* London: Macmillan.
74. Jerison, H. (1973) *Evolution of the Brain and Intelligence.* New York: Academic Press.
75. Jones, S. (1993) *The Language of the Genes.* London: Harper-Collins.
76. Jones, S., Martin, R., Pilbeam, D., & Bunney, S. (eds.) (1992) *The Cambridge Encyclopedia of Human Evolution.* Cambridge: Cambridge University Press.
77. Julesz, B. (1995) *Dialogues on Perception.* Cambridge, Mass.: MIT Press.
78. Jung, C. G. (1969) *Memories, Dreams, Reflections.* London: Fontana.
79. Kauffman, S. (1993) *The Origins of Order.* New York: Oxford University Press.
80. Kauffman, S. (1995) *At Home in the Universe.* New York: Oxford University Press.

81. Keller, H. (1902) *The Story of My Life*. New York: Doubleday.

82. Kelly, I. W. (1997) Modern astrology: a critique. *Psychological Reports*, **81**, 1035–66.

83. Kendrew, S. J. (ed.) (1994) *The Encyclopedia of Molecular Biology*. Oxford: Blackwell.

84. Kingdon, J. (1993) *Self-made Man and His Undoing*. London: Simon & Schuster.

85. Koertge, N. (1995) How feminism is now alienating women from science. *Skeptical Inquirer*, **19**, 42–3.

86. Koestler, A. (1972) *The Roots of Coincidence*. New York: Random House.

87. Krawczak, M., & Schmidtke, J. (1994) *DNA Fingerprinting*. Oxford: Bios Scientific Publishers.

88. Kurtz, P., & Madigan, T. J. (eds.) (1994) *Challenges to the Enlightenment*. Buffalo, New York: Prometheus Books.

89. Lamb, T., & Bourriau, J. (1995) *Colour: Art & Science*. Cambridge: Cambridge University Press.

90. Leakey, R. (1994) *The Origin of Humankind*. London: Weidenfeld & Nicolson.

91. Leakey, R., & Lewin, R. (1992) *Origins Reconsidered*. London: Little, Brown.

92. Leakey, R., & Lewin, R. (1996) *The Sixth Extinction*. London: Weidenfeld & Nicolson.

93. Lettvin, J. Y., Maturana, H. R., Pitts, W. H., & McCulloch, W. S. (1961) Two remarks on the visual system of the frog. In W. A. Rosenblith (ed.), *Sensory Communication*. Cambridge, Mass: MIT Press.

94. Lewis, C. S. (1939) Bluspels and Flalansferes. Chapter 7 of C. S. Lewis, *Rehabilitations and other Essays*. Oxford: Oxford University Press.

95. Lieberman, P. (1991) *Uniquely Human: The Evolution of Speech, Thought, and Selfless Behavior*. Cambridge, Mass: Harvard University Press.

96. Lofting, H. (1929) *Doctor Dolittle in the Moon.* London: Jonathan Cape.

97. Lovelock, J. E. (1979) *Gaia.* Oxford: Oxford University Press.

98. Margulis, L. (1981) *Symbiosis in Cell Evolution.* San Francisco: W. H. Freeman.

99. Margulis, L., & Sagan, D. (1987) *Microcosmos: Four Billion Years of Microbial Evolution.* London: Allen & Unwin.

100. Maynard Smith, J. (1972) The importance of the nervous system in the evolution of animal flight. In *On Evolution.* Edinburgh: Edinburgh University Press.

101. Maynard Smith, J. (1993) *The Theory of Evolution.* Cambridge: Cambridge University Press.

102. Maynard Smith, J. (1995) Genes, Memes, and Minds. *The New York Review of Books*, 30 November 1995, 46–8.

103. Medawar, P. B. (1982) *Pluto's Republic.* Oxford: Oxford University Press.

104. Medawar, P. B., & J. S. (1977) *The Life Science.* London: Wildwood House.

105. Medawar, P. B., & J. S. (1984) *Aristotle to Zoos.* London: Weidenfeld & Nicolson.

106. Miller, G. F. (1996) Political Peacocks. *Demos*, **10**, 9–11.

107. Minsky, M. (1985) *The Society of Mind.* New York: Simon & Schuster.

108. Mollon, J. (1995) Seeing colour. In T. Lamb & J. Bourriau (eds.), *Colour: Art and Science.* Cambridge: Cambridge University Press, 127–50.

109. Monod, J. (1970) *Chance and Necessity: An Essay on the National [sic] Philosophy of Modern Biology.* Glasgow: Fontana.

110. Morris, D. (1979) *Animal Days.* New York: William Morrow & Co.

111. Muller, R. (1988) *Nemesis: The Death Star.* London: William Heinemann.

112. Myhrvold, N. (1998) Nathan's Law (interview with Lance Knobel). *Worldlink*, World Economic Forum, 17–20.

113. Nesse, R., & Williams, G. C. (1994) *Evolution and Healing: The New Science of Darwinian Medicine*. London: Weidenfeld & Nicolson.

114. Partington, A. (ed.) (1992) *The Oxford Dictionary of Quotations*. Oxford: Oxford University Press.

115. Peierls, R. E. (1956) *The Laws of Nature*. New York: Scribners.

116. Penrose, A. P. D. (ed.) (1927) *The Autobiography and Memoirs of Benjamin Robert Haydon, 1786–1846*. London: G. Bell.

117. Penrose, R. (1990) *The Emperor's New Mind*. London: Vintage.

118. Pinker, S. (1994) *The Language Instinct*. London: Viking.

119. Pinker, S. (1997) *How The Mind Works*. London: Allen Lane.

120. Polkinghorne, J. C. (1984) *The Quantum World*. Harlow: Longman.

121. Randi, J. (1982) *Flim-Flam*. Buffalo, NY: Prometheus Books.

122. Rees, M. (1997) *Before the Beginning*. London: Simon & Schuster.

123. Rheingold, H. (1991) *Virtual Reality*. London: Secker & Warburg.

124. Ridley, M. (1996) *Evolution*. Oxford: Blackwell.

125. Ridley, M. (1996) *The Origins of Virtue*. London: Viking.

126. Rothschild, M., & Clay, T. (1952) *Fleas, Flukes and Cuckoos*. London: Collins.

127. Sagan, C. (1980) *Cosmos*. London: Macdonald.

128. Sagan, C. (1995) *Pale Blue Dot*. London: Headline.

129. Sagan, C. (1996) *The Demon-Haunted World*. New York: Random House.

130. Sagan, C., & Druyan, A. (1992) *Shadows of Forgotten Ancestors*. New York: Random House.

131. Scott, A. (1991) *Basic Nature*. Oxford: Basil Blackwell.

132. Shermer, M. (1997) *Why People Believe Weird Things*. New York: Freeman.

133. Singer, C. (1931) *A Short History of Biology*. Oxford: Clarendon Press.

134. Smith, D. C. (1979) From extracellular to intracellular: the

establishment of a symbiosis. In M. H. Richmond & D. C. Smith (eds.), *The Cell as a Habitat.* London: The Royal Society of London.

135. Smolin, L. (1997) *The Life of the Cosmos.* London: Weidenfeld & Nicolson.

136. Snow, C. P. (1959) *The Two Cultures and A Second Look.* Cambridge: Cambridge University Press.

137. Sokal, A., & Bricmont, J. (1998) *Intellectual Impostures.* London: Profile Books.

138. Stannard, R. (1989) *The Time and Space of Uncle Albert.* London: Faber & Faber.

139. Stenger, V. J. (1990) *Physics and Psychics.* Buffalo, NY: Prometheus Books.

140. Storr, A. (1996) *Feet of Clay: A Study of Gurus.* London: HarperCollins.

141. Sutherland, S. (1992) *Irrationality: The Enemy Within.* London: Constable.

142. Thomas, J. M. (1991) *Michael Faraday and the Royal Institution.* Bristol: Adam Hilger.

143. Tiger, L. (1979) *Optimism: The Biology of Hope.* New York: Simon & Schuster.

144. Twain, M. (1876) A literary nightmare. *Atlantic*, January.

145. Vermeij, G. J. (1987) *Evolution and Escalation: An Ecological History of Life.* Princeton: Princeton University Press.

146. von Holst, E. (1973) *The Behavioural Physiology of Animals and Man: The Selected Papers of Erich von Holst.* London: Methuen.

147. Vyse, S. A. (1997) *Believing in Magic: The Psychology of Superstition.* New York: Oxford University Press.

148. Watson, J. D. (1968) *The Double Helix.* New York: Atheneum.

149. Weinberg, S. (1993) *Dreams of a Final Theory.* London: Vintage.

150. Whelan, R. (1997) *The Book of Rainbows: Art, Literature, Science, and Mythology.* Cobb, Calif.: First Glance Books.

151. White, M., & Gribbin, J. (1993) *Einstein: A Life In Science*. London: Simon & Schuster.

152. Williams, G. C. (1996) *Plan and Purpose in Nature*. London: Weidenfeld & Nicolson.

153. Wills, C. (1993) *The Runaway Brain*. New York: Basic Books.

154. Wilson, E. O. (1998) *Consilience*. New York: Alfred A. Knopf.

155. Wolpert, L. (1992) *The Unnatural Nature of Science*. London: Faber & Faber.

156. Yeats, W. B. (1950) *Collected Poems*. London: Macmillan.

INDEX

Numbers in brackets refer to relevant works numbered in the bibliography.

Abraham 217
Adams, Douglas 29
aeroplanes, scepticism about 130–31, (20)
lack of scepticism about 138–9
Afrikaners, genetic disease 104, (11)
Akenside, Mark 38
aliens *see* life, extra-terrestrial
altruism, individual and gene selfishness 212, (31, 125)
angler fish 174–6, 245–6
animal as model of world 240
anteaters 242
ants 252–3, (68)
Appleyard, Bryan 37, (2)
aquatic mammals 242–5
arms race 232–3, (30, 145)
arts, spending on 5–6
Asatru Folk Assembly 19
Asimov, Isaac 27, 118, 142, (3)
asteroids, and mass extinctions 76, (1)
astrology 115–24, 185–6, (37)
astronomy 115–18, (122, 127, 128)
At Home in the Universe (S. Kauffman) 202–3, 207, (80)
Atkins, Peter, ix, 18, 30, (4)
swimming analogy for refraction, 44–5, (5)
Attneave, Fred 259, (6, 8)
Auden, W. H. 15, 199
aunt, Apthorpe's or levitating 131
Maud 33
Australopithecus 288, 289, (90)
Aztecs 181, (54)

bacteria 9, (99)
and Gaia hypothesis 223–4, (97)
luminous 228
organelles 226–8,
spirochaete 228, 230
bad poetry in science 180, 187, (58, 59)
barcodes x–xi, 49, 71, 81–2, 102
Barlow, Horace 257–8, (8)
bats, and sound 70, 72
and model of world 283, (31)
beauty 63–4
Bede, the Venerable 3
Bellamy, David 124
bestseller lists 291, (30)
Beyond Belief (television series) 126–7

Big Bang theory 60, (122, 135)
birds
flight control 274, (100)
keeping visual world constant 280
opening milk bottles 305, (52)
pigeons in Skinner box 162–5
sex chromosomes 236
song 79–81
birthdays, coincidence 152–4
Blackmore, Susan 302, 308, (10)
Blake, William 16–17
Blind Watchmaker, The (R. Dawkins) 292, 293, (30)
blood, resembling seawater? 255
bombs, atomic 290–91
bonobos 211–12
Book of Rainbows (R. Whelan) 47, (150)
Bormann, Martin 92
Bragg, Melvyn 32, (12)
brain
evolution 286–313, (74, 153)
as on-board computer 286–7
protected from redundant information 259–65, (6, 8)
branch point explosion hypothesis 201–2, 208, 209
Bricmont, Jean 41, (137)
Brockman, John xiii, 189, (13, 14)
Brooke, M. 246, 249, 250–51, 252, (27)
Browne, Sir Thomas, *Urne Buriall* 14
bug detectors 263–4, (93)
Burnell, Jocelyn Bell 34
butterflies 245
Cain, A. J. 254, (53)
Calvin, William 8, 299–300, 301, (16, 17)
Cambrian, evolution in the 200–206, 208, 209, (21, 59)
explosion 201, 204
camouflage, 240
Carey, John 35–6, (18)
Cartmill, Matt 20, (19)
cat, Cheshire 227, 307, (134)
caterpillars 140, 172–4, 245
CDs (laser discs) 79
cell, complicated structure 9
chain reactions 290–91
chance, exploited by psychics 145–7
Chandrasekhar, Subrahmanyan 63, (122)
change, sensory signalling of 260, 264, (8, 93)
chaos theory, and popular culture 188
Chardin, Teilhard de 184–5, 186, (103)
Chaucer, Geoffrey 251
children, credulity in 138–44, (147)
chimpanzees 297, 305, (35)
pygmy (bonobos) 211–12
chloroplasts 227–8, (98)
chromosomes, (11)
sex 236
tandem repeat regions 98–100, (87)
cilia 228, 230, (98)
Clarke, Arthur C. 27, 129, 130
his Third Law 129, 132, (20)
Climbing Mount Improbable (R. Dawkins) 305, (33)

codebooks 265
co-evolution 231–3, (42)
  co-adaptation 231, 232, 233, 289–90, 291–2
  self-feeding co-evolution of the human brain 289, 292, 294–312, (153)
coin theft 105–6
coin tossing trick 145–6
coincidences 145–60, 176–8
Coleridge, Samuel Taylor 40–41, 47–8, (56)
Color for Philosophers: Unweaving the Rainbow (C. L. Hardin) 58n
colour 44, (89)
  colour vision 53–8
combination lock 148
comets, hitting Earth 76–7, (1, 111)
communication, by early humans 296, 297–9
computers
  advances in 292–4, (48, 112)
  compared to the brain 286–7, 288, 289
Comte, Auguste 51
conception 1–2
conjurors 127, 128–9
Consciousness Explained (D. C. Dennett) 307, (38)
consonants, 78
conspiracy theory meme 304, (39)
constellations 116
Conway Morris, Simon 208, (21)
cooperation, genetic 213–33, 238
'copy me' 305, (33)
cosmic rays 52
Cottrell, Sir Alan 32
crabs, hermit 241
Creation Revisited (P. Atkins) 44, (5)
creationists, 'secular' 21, (19, 45)
creatures, fabulous 136–7, 210, 212
Crick, Francis 89–90, 191, 269, (24, 148)
cricket song 69
critical mass 290, 291
Cromwell, Oliver, his bladder 179
  his warts 268–9
Crucible of Creation, The (S. Conway Morris) 208, (21)
crucifixion 182–3, (54)
cuckoos 246–52, (27)
culture, (42)
  relativism in science 18–20, 21, (63)
  stereotypes 119–20
cycles, long-wavelength 74–7

da Vinci, Leonardo 47
Daily Mail newspaper, panders to astrology 115
Dalrymple, Theodore 108–9
Darwin, Charles, *On the Origin of Species* 16, (26)
Darwin, Erasmus 18
Darwin's Dangerous Idea (D. C. Dennett) 208, 303, (39)
Davies, N. 246, 249, 250–51, 252, (27)
Davy, Sir Humphry 40, (142)
de Waal, Frans 211, 212, (35)
Deacon, Terrence 309–10, (36)
Dean, G. 122, (37)
dendrochronology 82

Dennett, D. C. 207, 208, 283, 302, 303, 306–7, (38, 39)
Deutsch, David 50, (40)
Diamond, Marian C. 286
Diana, Princess of Wales 115, 125
Dickinson, Emily 66
dictators, moustaches 87
diversity, evolution of 200–201, (101, 124)
DNA
 amounts in different species 97
 helical structure 187
 junk 97–8, (87)
 of parasites 226–7, (29)
 probe 100, (83, 87)
 a reflection of ancestral worlds 239, 254–5
 and Rosalind Franklin 191, (148)
 selfish/ultraselfish 98, (31)
 symbolic meaning, 183
 tandem repeats 98–100, (83, 87)
 'you believe in . . .?' 190, (45)
 see also genes
DNA fingerprinting 83–92, 112–13, (87)
 background to the technique 95–100
 a national DNA database 109–12
 objections to evidence using 92–109
 single-locus 102–3
 and statistics 90–91, 94–5, 103–9
 technique 100–103
Doctor Dolittle 53, (96)
dominance hierarchies 237, 238
Doppler, Christian 59
Doppler shift 59–60, 62
Doyle, Sir Arthur Conan 136, (121)
dreams 158, 282
drugs, birdsong acting as 79–81
Druyan, Ann 114, (130)
'dumbing down' 21–4
Dunnock Behaviour and Social Evolution (N. B. Davies) 246, (27)
dunnocks 246, 248–9, 251–2

ears 66, 68–9
 see also hearing
earth
 and the Gaia hypothesis 222–4
 lucky to be alive on 4–5
Eddington, Sir Arthur 42, 111, 135, (44)
edges, sensory signalling of 262
Edinburgh, Prince Philip Duke of, 91
Ehrenreich, Barbara 190, (45)
Einstein, Albert 42, (46, 138)
Eldredge, Niles 195, 196
electromagnetic spectrum 52–3, 58–9
 see also light
electrophoresis, gel 101–2
elements, and Fraunhofer lines 50
elephant, swinging penis 73–4
Ellsworth, Phoebe 190, (45)
Elton, Charles 74
emotion, and poetry 79
endoscopy 272–4
enzymes 54, 55
 restriction enzymes 100–101, (87)

errors, false negative and false positive (type 1 and type 2) 94, 171–7
eternal metaphors of paleontology 193–9, (58)
Evans, Christopher 288, (48)
evolution (101, 124)
  and bad poetic science 192–209
  and catastrophism 198
  general evolutionism 192–3
  gradual versus episodic 195–200
  of the human brain 286–313, (153)
  opposition to 20–21
  'top down' theory 203–6
  see also co-evolution
exorcist, enterprising but fraudulent 121
'experience' of genes 236–8
experimental design 167–8
'experimentum crucis' (Newton) 43
Extended Phenotype, The (R. Dawkins) 308, (29)
extinctions, mass 75–7, 195, 199, 222–3
Eye and Brain (R. Gregory) 276, (61)
eye muscle paralysis experiment (Kornmuller) 281, (146)
eye-witnesses, evidence from 85–6
eyes, and colour vision 54, 55–6, (89)

Faber Book of Science, The 35, (18)
Fabric of Reality, The (D. Deutsch) 50, (40)
faces
  eagerness of the brain to construct 266–9
  recognition 87–8, 258–9, 266
fairies 136, (121)
'false colour' images 57
familiarity, anaesthetic of 6
Faraday, Michael 5, 23, 186–7, (142)
Father Christmas 141
Fatima, Our Lady of 134–5
feet, webbed 242, 244–5
feminism, and bad poetic science 189–92, (63, 85)
Feynman, Richard 41–2, 50, 150–51, (50, 51)
fire, discovered by humans 11
fish
  angler 174–5, 245–6
  and luminous bacteria 228
flagella, bacteria 228
flies
  creating a visual illusion on 281
  mimicking jumping spiders 290, 291–2
flight 274, 275, (100)
flying saucers 137–8
footprints, reading 297–9
Ford, E. B. 214–15, (53)
foresight 302, 312–13
formants 78
fossils, age of 9–14
Fourier analysis 72–3, 76
Franklin, Rosalind 191, (148)
Fraunhofer lines 49–51, 59
Frazer, Sir James 180–81, 182, 183, (54)
Freeman, Derek 211n, (55)

frequency dependent selection 96
frogs, bug detectors 263–4, (93)
Frost, Sir David 126, 127
fun 22–4
furs, and predator–prey cycles 74

Gaia hypothesis 222–4, (29, 97)
galaxies, light from 59–60
gamblers, irrational 165–6, (141)
gel electrophoresis 101–2, (87)
gender or sex 246, (118)
gene pools 221, 254, 256
  learning from 'experience' 238–9
genes
  climate 214, 219
  cooperative 213
  describing ancient worlds 235–56
  isolated 90
  segregation distorter genes 218
  selfish genes are also cooperative 210–33, 238, (31)
  surviving in virtual worlds 284–5
  'switch' 215–16
  variation 95–6
  see also DNA
gentes 247, (27)
geological time, magnitude of 9–14
Gill, A. A. 34
God, good Darwinian 217
Golden Bough, The (J. Frazer) 180–81, (54)
Goldschmidt, Richard 196
Goldsmith, Edward 222–3
Good, I. J. 108, (57)
Gould, Stephen Jay 193–203, 207, (58,59)
gradualism 198–9, (30, 32, 33)
Grand Canyon 13–14
grandmother, Lettvin's 258
Graphical User Interface (GUI) 293–4, (112)
Gregory, Richard 23, 276, (61, 62)
Gross, Paul 20, 191, (63)
gulls, gliding 274

Haldane, J. B. S. 193
hallucinations 282
handwriting 88
  sex-related 166–70
Hardin, C. L. 58n, (65)
Harding, Sandra 191, (63)
harems, seal 238
harmonics 71
Haydon, Benjamin 38–9, (116)
hearing 66–72
Heaven's Gate cult 28
herbivores 220
Higher Superstition (Gross and Levitt) 191, (63)
history, and our existence 2–3
Hitler, Adolf 63, 112
Hobbes, Thomas 186
hollow mask illusion 267–9
Holst, Erich von 280–81, (146)
homeopathic magic 181, (54)
Homo spp.
  increase in brain size 287–8, 289
  language and communication 296, 297
  see also humans
homunculus 283, (38)

horoscopes 120, 123–4
How the Mind Works (S. Pinker) 179, 184, 191, 278, (119)
Hubbard, L. Ron 27–8
Hubble, Edwin 59, 60
Hubble's law 60
Human Genome Project 90, 94, (11)
humans
evolution of the brain 286–313, (153)
impressed by coincidences 177–8
see also Homo spp.
Hume, David 133–5, (71, 139)
Humphrey, Nicholas 152, (71, 72)
Huxley, T. H. 179, (73)
hybrids 214–16

identity parades 86–7, 94
illusion 276, (61)
imagination 312–13
imitation, by individuals 305–6, 308
Independent newspaper 107
India, nuclear tests by 30n
information, technical meaning of 259
infrared rays 52
insects
ears 68–9
social 252–3
and virtual reality 280–81
insurance companies 111
Intellectual Impostures (Sokal and Bricmont) 41, (137)
intuition 176–7, 178–9
iris-scanning machines 88–9

James, William 175
Jeffreys, Alec 99
Jenkins, Simon 36–7
Jesuits 144
Jesus, face of, in dishcloth 267
Julesz Effect 278, (77, 119)
Jung, C. G. 154–5, (78)
Jupiter 62
jurors, educated in science 83–5

Kammerer, Paul 157, (86)
Kauffman, Stuart 202–3, 204, 207, (79, 80)
Keats, John x, xiii, 26, 27, 38–9, 41, 61, 64, 79, 81, 145, 313, (22)
Keller, Helen 257, (81)
Kelvin, William Thomson, Lord 129–130, (20)
Kennewick Man 18–20
keyhole, visual 258–9
Kimura, Motoo 95, (124)
Kingdon, Jonathan 306, (84)
Koertge, Noretta 189–90, (85)
Koestler, Arthur 154, 157, (86)
Krebs cycle 9
Kropotkin, Peter 211

Lamb, Charles 38–9, 114, (116)
language 294–6, 309–10, (36, 95, 118, 119)
Lanier, Jaron 272
laser discs 79
lateral inhibition 262, (8)
launch windows 301, (16)
law
and DNA evidence 83–113
and eye-witness evidence 85–6

Lawrence, D. H. 25, 51
lawyers, woe unto 83
Leakey, Richard 205–7, 242, (90, 91, 92)
Leonardo da Vinci 47
Lettvin, J. Y. 258–9, 263–4, (93)
Levin, Bernard 31–2
Levin, Margarita 191–2, (63)
Levitt, Norman 191, (63)
Lewin, Roger 205–7, (91, 92)
Lewis, C. S. 186, 310, (94)
life
  extra-terrestrial 60–63, 90, 117–18, 137–8
  and geological time 9–14
  and personal identity 1–5
  science and the wonder of 5–9
light
  divisibility and the rainbow 39, 40, 42–9
  and the electromagnetic spectrum 52
  red/blue-shifted 59–60, 62
  refractive index 44
  velocity 44
  wave theory of 43
line detectors, and vision 263
line-ups (identity parades) 86–7, 94
longitudinal transmission 226
Lorenz, Konrad 31
Lovelock, James 223, 225, (97)
lucky to be alive 1–5
Lucy 288, (90, 91)

MacCready, Paul 275
McIntosh, Janet 190, (45)
macromutations 195–6, 197
mad cow disease 54
Magic Eye illusions 278, (119)
magical customs 180–82, (54)
Magnetic Resonance Imaging 59
mammals
  DNA reflecting ancestral environments 254–5
  water-dwelling 242–5
maps, and evolution of the human brain 297–9
Margulis, Lynn 224, 225, 226, 228, (98, 99)
marine chemistry, genes reflecting 255–6
Maynard Smith, John 207–8, 222–3, 275, (100, 101, 102)
Mead, Margaret 211, 211n, (55)
Medawar, P. B. 30, 32–3, 184–5, 312, (103, 104, 105)
membranes, in cells 9
Meme Machine, The (S. Blackmore) 308, (10)
memeplexes 306, (10)
memes 302–9, (10, 31, 38, 39)
memory 257–8
Mengele, Josef 91–2, 112
metaphors, and evolution of the human brain 310–12
methane 223
mice, genes 218, 254
microwaves 52
Minsky, M. 309, (107)
miracles 133–4, (71, 129)
mites 241
mitochondria 9, 225–6, 227, (98, 134)
Mixotricha in termite gut 229–30, (98)

model of world 277, (23)
animal as 240
molecular clock 208–9, (62, 83)
moles 241
Mollon, John 58, (108)
monkeys, colour vision 58, (108)
Monod, Jacques 187, (109)
Montgomery, Field Marshal, agrees with God 217
moon 51
rhythmic influences of 75
Moore's law 288, 292, 294, (112)
morning, why get up in the ix, 6
Morris, Desmond 2, (110)
moths, Lesser Yellow Underwing 215–16, (53)
mouse 254
mouse, computer 293
moustaches 87
music 36, 70–72
mutations 195–6, 197, 235–6
and DNA fingerprinting 95–6, 98
Myhrvold, Nathan 294, (112)
mysticism 17, (71, 129)

Native Americans, and Kennewick Man 18–20
natural selection 245, (30, 33, 101, 124)
and genes 216–17, 221, 233
Necker cube 276, (61)
Nehru, Jawaharlal 30
Nemesis (hypothetical star) 77, (111)
Neptune, discovery 61, (122)
nerve impulses, and colour vision 55–6, (89)
neutral theory of evolution 95, 195, (124)
New York Review of Books (J. Maynard Smith) 207–8, (102)
Newcomb, Simon 130, (20)
Newton, Sir Isaac x, 191, 313
his prism 39–40, 42–3, (89)
nightingales 79–80
Norsemen, and Kennewick Man 19

On Giants' Shoulders (M. Bragg) 32, (12)
Oort cloud 77, (111)
Opportunities for coincidence 159, 178
Optimal Foraging Theory 163
optomotor apparatus 280
O'Reilly, John Boyle 15
Origin of Species, On the (C. Darwin) 16, (26)

p-values 170–71, 176
Pale Blue Dot (C. Sagan) 114, (128)
paranormalism xi–xii, 124–9, (37, 71, 121, 129, 139)
and credulity 138–44, (141, 147)
parasites 226, 241, (29, 113)
parentage, and DNA fingerprinting 91, 112–13
Parsonstown, Leviathan of (telescope) 26
paternity *see* parentage
patterns, real and imaginary 160–62
PCR (polymerase chain reaction) 92–3, (87)
Peake, Mervyn 1

penny, diary of a 235
perpetual motion 135
personality, and astrology 118–19
PETWHAC, defined 151
photons 43
Physics and Psychics (V. Stenger) 188, (139)
Pickering, William Henry 130–31, (20)
pigeons, in a Skinner box 162–5, 172, (28)
Pinker, Steven 179, 184, 191, 278, (118,119)
pitch 69
planets
  discovery 61
  outside our solar system 61–2
Ples, Mrs 288
plumber, virtual 273
Pluto, discovery 61
poetry
  bad poetic science 180–209
  and emotion 79
  and science 15–18, 24–7
polymerase chain reaction (PCR) 92–3, (83, 87)
polymorphism 253–4, (11, 53)
porphyria variegata 104, (11)
post-modernism 41, (63, 137)
predator–prey population cycles 74
Preminger, Otto 118
Presley, Elvis 125
pressure 66–7
Principal of Least Action 44, (5)
prions 54–5
prisms 42–3
prosopagnosia 269, (24)
proteins, (83)
  alternative forms 96
  prions 54–5
psychic phenomena, explained away 145–7, 149–52, (71, 129, 139)
psychology, evolution of 177, 179, (7, 125)
pterosaurs, flight 275, (100)
pulsars 34, (122)
Python, Monty 119

quantum theory 50, (15, 40, 117, 120, 131, 149)
  and light 43
  misuse of 188
quarks, and Bernard Levin 31–2

radio waves 52
rainbows 45–9, (150)
Raine, Kathleen 6
rainforests, and gene survival 221, 222
rainmaking 161, 181–2, (54)
Randi, James 123, 128, (121)
recognition, sensory 258–65
redundant sensory information 259–65, (6, 8)
refraction, refractive index 44, 45
reincarnation 125–6
reliability test, 122, (37)
religious customs 180–83, (54)
restriction enzyme 100, (83, 87)
retinal cells 55, 56
Rettenmeyer, C. W. 241
Romanov family, identification 91
Rothschild, Miriam 241, (126)
rowing analogy 216, (31)

Rowland, Ian 128
Royal Institution Christmas
Lectures 23, 145, 292, (142)
Ruskin, John 49

Sagan, Carl x, 18, 30, 63, 114, 138,
(127, 128, 129, 130)
Sagan, Dorion 224, (99)
satellite pictures, 'false colour' 57
scapegoats 182–3, (54)
scepticism 129–38, 142, 143, (71,
121, 129, 139, 147)
science (40, 41, 51, 60, 88, 103, 129,
149, 154)
bad poetic science 180–209,
(58, 59)
and culture 18–20, 21, (136,
154)
'dumbing down' of 21–4
poetic, and the sense of wonder
15–37, 41–2, 63, (128, 129)
and potential jurors 83–5
society's perception of 29–37,
(34, 41, 155)
use of 5–6
versus intuition 178–9, (155)
science fiction 27–9, 37
Scientology 27–8
seals 238
self-feeding processes 289–4
and evolution of the human
brain 290–312
Selfish Gene, The (R. Dawkins)
212–13, 216, (31)
sex, and genes 236, 245–6
sex chromosomes 236, 247, (11)
sexual selection 292, 309, (25)
Shakespeare, W. 180, 251, 252
Sheppard, P. M. 254, (53)
Shermer, Michael xi, 188, (132)
Simonyi, Charles xii–xiii, 51
sine waves 69–70
Singer, Charles 89–90, (133)
Sitwell, Sacheverell 13
Sixth Extinction, The (Leakey and
Lewin) 205, (92)
skeletons, identification 91–2
Skinner boxes 162–5
slang 310–11
smell, recognizing people by 88,
89
Smith, Adam xii, 210, 212
Smith, D. C. 227, (134)
snails, polymorphic 253–4, (53)
snakes, learning dangers of 143
Snow, C. P. xiii, 115, (136)
Society of Mind, The (M. Minsky)
309, (107)
Soul Searching (N. Humphrey)
152, (71)
sound 66–73
birdsong 79–81
speech 77–9
virtual 261
South African Dutch, genetic
disease 104, (11)
Southern blot 102, (83, 87)
spectroscopy 51–2, (89)
species as 'averaging computer'
239, 256
species splitting 200, (124)
speech 77–9
Spencer, Herbert 192, (103)
spiders, flies mimicking 290,
291–2
spinning, and giddiness 279–80

spirochaetes 228, 230
Spooner, W. A. 87–8
squid, colour change 7–8
stars
binary 62, 77
birth year 117
study using spectroscopy 51–2
statistics 166–76
animals as practitioners of, 163, (28)
and coincidence 145–60, 176–8, (141)
and patterns in nature 160–63
statistical significance 170–71
and superstitious habits 163–6, (28)
Stenger, Victor 188, (139)
stereoscopy 277–8, (61, 77, 119)
stereotypes, cultural 119–20
subjective present 3, (69)
sun
classification 51–2
sister star to the 77, (111)
sunspots 74
Sunday Sport 124, 125
superstition 162, 163–6, (28, 71, 129, 147)
surgeon, virtual 272–3
swimming, adaptations to 242
swifts 240–41
symbiosis 229–31, (98, 134)
Symbolic Species, The (T. Deacon) 309–10, (36)
symbolism 181–4, (54)

tall stories 133
telepathy 127–8, (71, 121, 139)
termites 229–31, 252–3
theology 183
thermodynamics, second law 135, (4, 136)
thinking 8
Thomson, James 64–5
Thomson, William, lord Kelvin 129–30, (20)
throwing, and evolution of the human brain 299–302, (16)
time, (40, 122, 135)
beginning of 60
looking back in 116–17
and subjectivity 3–4, (15, 69)
vastness of geological 9–14
tits, blue, opening milk bottles 305, (52)
toad, eyes in roof of mouth 196, (33)
trade, British distaste for 32–3
transubstantiation 181
tree analogy for Cambrian error 205
tree rings 82
trilobites, age 9–12
tropical rainforest 221–2
truth 21, (88)
tunes, repeated in the mind 302–3, (39)
tuning forks 67–8
Twain, Mark 303–4, (144)

ultraviolet rays 52
undulipodia 228, (98)
unit of selection 199, 220, (29)
universe, expanding 60, (122, 135)
Unnatural Nature of Science, The (L. Wolpert) 178–9, (155)
Uranus 61

vaccination programmes 291
variation, needed for DNA fingerprinting 95–8
virtual reality 270–74, (123)
  model in the brain 261, 262, 266, 274–85
vision (61)
  colour 53–8
  and line detectors 263
  and redundant information 261–4
  see also virtual reality
voice, recognizing people by 88
vowels 77–8

War of the Worlds, The (radio dramatization) 137
Ward, Lalla xiv, 118, 154, 302
watch
  initials coincidentally engraved 154–7
  started by psychic forces? 152
  stopped by psychic forces? 147, 149–52
water dwelling mammals 242–5
Watson, James D. 89–90, 191, (148)
Watson, Thomas J. 131
Waugh, Evelyn 131
wave theory of light 43
waveforms, periodic 66–82
Weldon, Fay 33–4
Welsh language 57
Whelan, R., 47, (150)
Why People Believe Weird Things (M. Shermer) 188, (132)
Windows 293
winds 67
Winston, Robert 145
Wollaston, William 49
Wolpert, Lewis 30, 178–9, (155)
Wonderful Life (S. J. Gould) 200, 207, (59)
Wordsworth, William 38–40, 48

X-Files, The (television series) 28
X-rays 52
Xerox PARC 293

Yeats, W. B. 26, 27, 235, 299, (156)

Zipf's Law 264–5
'zoologist of the future' 240